THE SCIENCE
BEHIND
NOAH'S FLOOD

THE SCIENCE
BEHIND
NOAH'S FLOOD

by

Robert E. Farrell

Published by: **R. E. FARRELLBOOKS, LLC**
Sun City West, Arizona

2nd Edition – 2016
Hard Cover – November, 2021

R. E. FARRELLBOOKS, LLC
Sun City West, Arizona

www.alienlog.com

ISBN: 978-0-9963587-8-1 (Hardcover)
978-0-9759116-9-3 (Paperback)
978-0-9963587-0-5 (E-Book)

LCCN: 20xxxx

Cover design by Wendy Farrell plus several figures are by the author.

ACKNOWLEDGMENTS

This book never would have happened had it not been for the help of many people and the encouraging response to my lectures, ***Deluge: Noah's Flood*** and ***The Science Behind Noah's Flood,*** as well as letters from readers of the books in the ***Alien Log*** series of novels. Below is a partial list of credits.

First and foremost, much credit goes to my late wife, Linda, who stood by me patiently during the years of research and writing that went into these books. Thanks to my daughter, Wendy who helped in design the book covers and worked diligently to get my books to market.

I tip my hat to the late Zecharia Sitchin who got me interested in this subject. Thanks to NASA, Google Earth, and Wikipedia for their wonderful photographs. Cover design is an adaptation of the famous 1846 painting, ***Noah's Ark,*** by Edward Hicks.

I am deeply indebted to Jennifer Hope, Mesa Verde Media Services, for her tireless and superb efforts in copy editing.

In Loving Memory of
My
Wife and Son

TABLE OF CONTENTS

Introduction

In almost every country, one can find a deluge story. Many involve a deluge, a divinity, a hero, and saving humanity. In ***Global Flood Stories***, Mark Isaak has listed 262 stories (see www.talkorigins.org/faqs/flood-myths.html). Some are local flood stories and some are global events.

The focus of the following book is the sudden deluge or series of deluges which may have occurred over a period of months or years and are described in the well-known story of Noah's Flood as depicted in Christian, Jewish, Islamic, and Hindu traditions. The Chinese history also contains a Great Flood which occurred three thousand years ago. It lasted for at least two generations and thus was not a sudden event of short duration.

Many in the scientific community at the end of the nineteenth century considered the story of Noah's Flood to be a myth, pure fiction, a child's story to demonstrate God's power. Sir Leonard Woolley, the great British archeologist of that period, changed that notion in 1929 when he published ***Ur of the Chaldees: A Record of Seven Years of Excavations***. He was convinced that the ancient city of Ur about 200 miles southeast of present-day Baghdad was the biblical birthplace of Abraham. He began his dig there and uncovered evidence of a major flood. He believed that this flood was the flood mentioned in Genesis as Noah's Flood. He also was convinced that it was not a world flood but rather a flood which affected only the region of southern Mesopotamia between the Tigris and Euphrates rivers. While periodic flooding in this region was common, he felt this was a major flood affecting an area perhaps 400 miles long and 100 miles wide. To the inhabitants living in that region, this would have seemed to be a world flood.

Although Noah's story is considered by some, even today, to be a myth or legend, there is little doubt in this author's mind that it has a basis in fact. However, there are many questions to be answered. Was the Earth entirely covered by water? Did all humanity but a handful perish? Did it happen 4,268 years ago as would be determined from Genesis? What caused the flood to happen? These questions will be addressed in this book.

It is difficult to believe that the entire Earth was covered in water. There is no evidence for this. We do know that as continents rise and fall during movement of tectonic plates some land that is now dry was once the floor of an ocean. In the desert southwest of the United States, it is possible to find sea shells lying on the ground. Mark Isaak has pointed out that if the Earth had been completely covered by water, the polar ice caps would have been floated off their beds and broken up. Also, he points out that tree rings going back 10,000 years show no evidence of a world flood during that period. Consider this: if all of the water in the atmosphere were to precipitate out, it would not flood the entire Earth. Even though the Earth's atmosphere contains over 37 million billion gallons of water at any one time, it would only cover the entire surface (land and sea) of the Earth with an inch of rain. Finally, if the Earth were flooded to the extent mentioned in the Bible, where did all the water go when it subsided?

When did the Deluge happen? Various scholars, using chronology from the Bible, have placed the flood at about 4,268 years ago. At least one difficulty with that date is that some of the oldest pyramids in Egypt were built before 4,268 years ago. The Pyramid of Djoser is one of the world's oldest monumental structures constructed of dressed masonry and was built during the third dynasty about 4,620 years ago. Aside from 4,620 years of weathering, there is no indication it was ever submerged under thousands of feet of sea water. Also, the Egyptian records show no mention of that flood.

Saqqara pyramid of Djoser in Egypt. Credit: Charles J. Sharp from Wikimedia Commons

According to Noah's story every living thing was drowned except for Noah, his family, and the animals he saved. From the book of Genesis 6:17, God told Noah that he was going to destroy all flesh, "everything that is on Earth shall die." In Genesis 6:19, "And of every living thing, of all flesh, you shall bring two of every kind into the ark, to keep them alive with you; they shall be male and female." Thus every living thing on Earth came from Noah's Ark and multiplied to be the living planet we have today. That is hard to believe. As to what caused the flood and when, that will be addressed in the following chapters.

By convention, dates may be given by BP (before present), BC (before Christ), and BCE (before common era). To avoid confusion, this book will date events as BC meaning before Christ.

Chapter 1
Where did the Deluge occur?

The evidence points to Mesopotamia. This is the birthplace of civilization, and it is truly ironic that so much uncivilized activity is occurring, and has occurred, here in the late twentieth and early twenty-first centuries. Geographically, it is the area of the Tigris and Euphrates rivers and includes northeastern Syria; parts of Turkey, Iraq, Kuwait; and even sections of southwestern Iran. Within Mesopotamia, Iraq takes center stage in our story because it contained Sumer, Akkad, and Babylonia. Our story begins in Sumer where it all started.

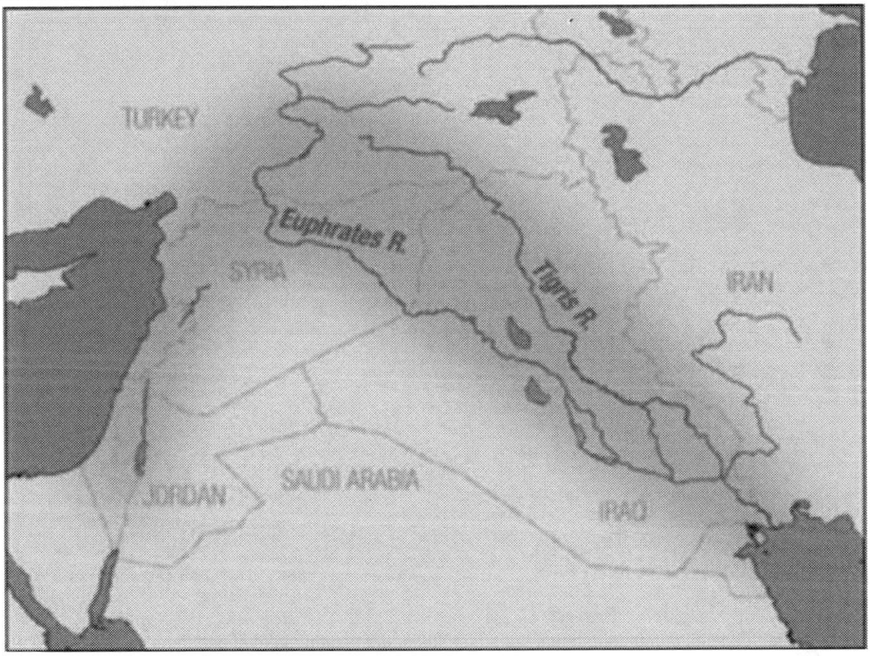

Map of Mesopotamia (shaded area between Tigris and Euphrates rivers). Credit: NASA

From this birthplace of civilization came many of the myths, traditions, laws, and inventions that were embraced by later civilizations. Civilization in Sumer was followed by Akkad and then Babylonia. In Sumer, Eridu was the first city and was followed by Ur. The map below shows the ancient civilizations.

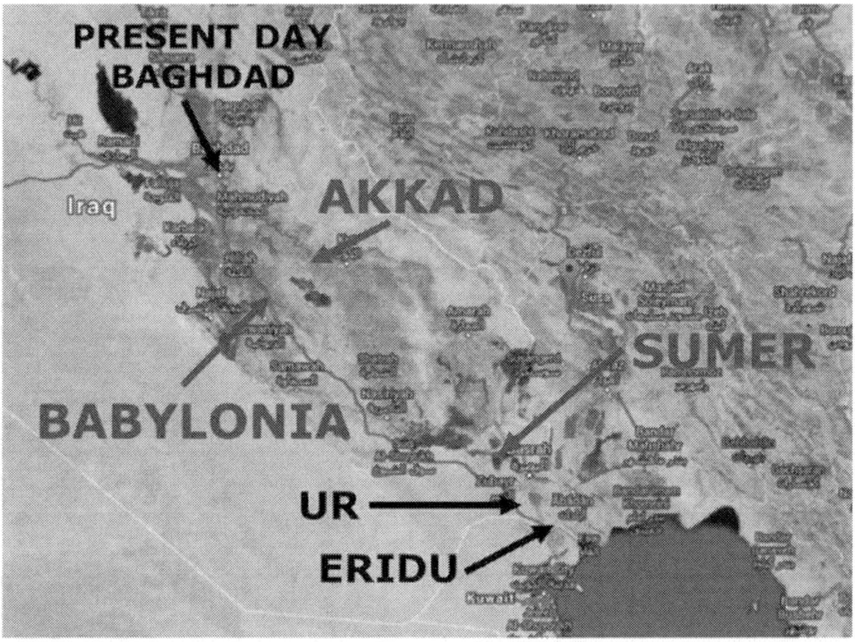

Author's depiction of Sumer, Akkad, and Babylonia as well as the first two cities in Sumer. Source map: Google Maps

The Sumerians were amazing in the sudden appearance of knowledge well beyond any other civilization before them. From a historical point of view, the invention of writing was most significant. It was their invention of cuneiform writing that was the beginning of written history. What we know of human activity before that is simply conjecture and is pre-history.

The list of Sumerian inventions is truly amazing. For example, seemingly overnight, they invented law, astronomy, astrology, medicine, and mathematics. They used a sexagesimal (base 60) number system. Today it is a superior system in modified form for some types of measurements such as for measuring time, angles, and geographic coordinates.

The number twelve was a sacred number for the Sumerians. Their pantheon consisted of twelve gods. Later civilizations follow suit. A day was two sets of twelve hours. The importance of twelve has carried forward to present day. We have twelve inches per foot, twelve months in a year, and twelve things constitute a dozen.

The Sumerians lived in a land that was (and still is) rich in oil. They invented ways to refine oil into products such as oil to light their lamps and pitch which was an important building material. They learned to mine and refine metals and produce alloys. They invented the shekel as a currency used in their trade with other states. They invented improved ways to farm and irrigate their fields. Their textile products were prized by others in the region. And finally, where would we be if the Sumerians had not invented the wheel?

Cylinder seal and clay "print." Credit: Jastrow, The Louvre (Public domain)

Cuneiform tablet from the Kirkor Minassian collection. c. twenty-fourth century BC. Credit: Library of Congress

(Public domain)

The high level of Sumerian civilization is best demonstrated by their art. Below are some examples of artifacts Sir Leonard Woolley recovered from the royal grave, known as the Great Death Pit, near Ur. While there is no evidence of human sacrifice outside of the royal court, it seems to have been a common practice to have the royal court commit suicide in the grave pit and be buried with the king or queen. The burial process would include burying musical instruments, the royal guards, chariots, and even the oxen who pulled the chariots. Below are examples of Sumerian art found in the grave pit of Queen Puabi.

Reconstructed Sumerian headgear necklaces found in the tomb of Puabi, housed at the British Museum. Credit: JMiall (Wikimedia Commons)

Ram in a Thicket. Credit: Jack1956 (Wikimedia Commons)

Chapter 2
The Sumerians per Zecharia Sitchin

Zecharia Sitchin. © Z SITCHIN www.sitchin.com

A wonderful series of books about the Sumerians was written by the late Zecharia Sitchin. His book, ***The 12th Planet: Book I of the Earth Chronicles***, reveals that the Sumerian civilization had a detailed understanding of our solar system as well as how Homo sapiens were created. This book was his first book of what he called the Earth Chronicles and is a documentary about the origin of the Earth and humans, man's celestial ancestors, and the Deluge.

At the beginning of his book, Sitchin explains that he has compared translations of early Hebrew texts with those of the Sumerian and Akkadian texts to create his rendering. His work is important to the topic of Noah's Flood because of the insight it gives to the earliest accountings of that period of time. His work is so important that I considered using a sub-title: ***The Redemption of Zecharia Sitchin*** for this book. I will spend considerable time in this chapter discussing Sitchin's thesis on the Sumerians and defending his findings in light of forty years of scientific discovery which has occurred since his book's publication in 1976. It was Sitchin's work and his hypothesis about Noah's Flood that became the seed that germinated into this book.

Sitchin was born in 1920 in Azerbaijan, SSR, and raised in Palestine. After earning a degree in economics from the University of London, he became a journalist and editor in Israel. In 1952 he moved to New York to work for a shipping company. During his early life, he began questioning the misinterpretations of the original Hebrew texts. He rejected the dogma being taught about stories in the Old Testament and wanted a better understanding of his own Hebrew religion. He believed many of the stories had their roots with Abraham, whom he believed came from the southern Sumerian city of Ur.

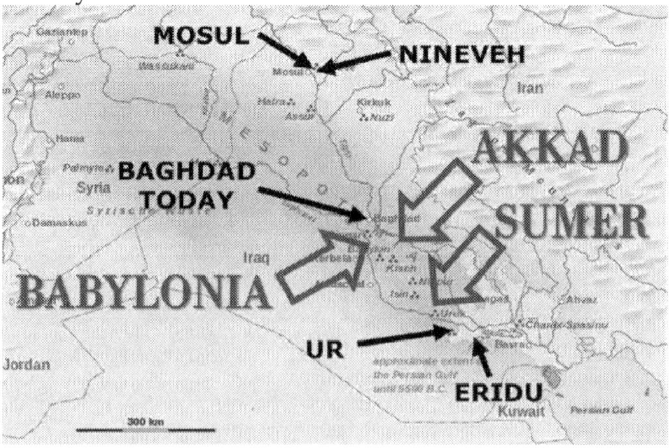

Author's depiction of Sumer, Akkad, and Babylonia as well as the first two cities in Sumer. Source map: Google Maps

Most of the history of that region is recorded in the cuneiform writings that appeared on thousands of ancient clay tablets discovered in the mid-1800s in Nineveh. This is an ancient town in northern Iraq near the Kurdish town of Mosul. These tablets had been part of the great library of Assurbanipal, who had been king of Assyria from 669 to 633 BC. Over the centuries, the library became lost and it was not until the mid-1800s that it was rediscovered. Thousands of clay tablets were hauled off to various museums around the world including the Baghdad Museum, the British Museum, the Museum of Ancient Near Eastern Antiquities in Munich, Germany, and the Near East Section at the University of Pennsylvania Museum of Archaeology and Anthropology.

Sitchin taught himself to translate Sumerian cuneiform and was fortunate to have the opportunity to visit some of the libraries mentioned above and several archeological sites. As we continue, it is important to remember that he published his book in 1976.

Since he had no degrees in antiquity, many from the mainstream scientific community consider his translations flawed. However, since publishing ***The 12th Planet*** in 1976, many of his interpretations have been proven by mainstream science to be accurate. The following are examples.

ASTRONOMY

In my first example, Sitchin describes the Sumerians as knowing that there were planets beyond those visible by the unaided eye. One has to wonder how they knew about them. Sitchin goes on to explained that the Sumerians believed a rogue planet with moons entered our solar system and was responsible for altering our solar system into what we see today. They called it Nibiru (Marduk in Babylonian). The fact that a rogue planet with moons could form and move freely between stars was only speculated and not known to astronomers in 1976.

Astronomers know that stars from when the gas and dust in a region of space collapses into a central region due to gravity. Pressure and temperature increase until conditions are right to initiate nuclear fusion of hydrogen into helium. This requires a core temperature of 14 million kelvin (about 25 million °F). However, it was known that if a planetary nebula's mass is less than 13 times the mass of Jupiter, the resulting body would not form sufficient pressure and temperature as it collapses to cause fusion. The body would become a brown dwarf or perhaps a rogue planet. If the planetary nebula is rotating due to turbulence, then as it collapses, a planetary disk will form out of which moons will form. This is the same process that occurs as stars and their planets are formed. In October of 2000, using new infrared telescopes, 18 free-floating rogue planets were discovered near Sigma Orionis, 1,200 light years away. They were able to be seen because of improvements in our infrared telescopes. Astronomers speculate they are less than 5 million years old and thus still hot enough to be seen in the infrared. More recently, a free-floating rogue planet, CFBDSIR 2149-0403, was discovered only 160 light years away. Astronomers speculate it is a young and cold planet between four and seven times the size of Jupiter. They think it could have formed from collapsing clouds of gas and dust just as does a star or brown dwarf. With time, astronomers may learn that the moons of such rogue planets may be able to support life just as Europa, a moon of Jupiter and Enceladus, a moon of Saturn. Both moons are thought to have life supporting conditions as a result of heating from tidal friction as they orbit their planets. If the nebula is rotating as it collapses, angular momentum is conserved and a planetary disk forms and planets accrete from this gas and dust. These planets revolve around the star in the same direction and in the same plane as the gas and dust. In our solar system, this plane is called the ecliptic. Just as planets accreted in our solar system, moons can accrete from a disk of gas and dust around a rogue planet.

Thus, it is possible to have rogue planets with moons wander the galaxy until they are captured by a nearby star. When they are captured, it is unlikely that they would fall into an orbit that lies in the same plane (ecliptic) as the star's natural planets. Its orbit would probably be inclined to the ecliptic. Also, its orbit would be an elongated ellipse and there is a 50 percent chance it would orbit in the opposite direction (retrograde) to the natural planets.

The Sumerians believed Nibiru with its moons were captured by our sun about four billion years ago. This was about half a billion years after our solar system first formed and at a time when many planets were still hot and plastic. According to Sitchin's interpretation of the ancient Sumerian myths, Nibiru entered an orbit with its closest approach to our sun being where the present asteroid belt exists today at about 2.7 astronomical units (AU). (One astronomical unit is the distance between the Earth and the sun.) Because of the timing of human events, Sitchin concluded that the orbital period for Nibiru was 3,600 years. Assuming Nibiru's closest approach to the sun is about 3 AU, a simple calculation using Kepler's Laws reveals that Nibiru's farthest distance from our sun is about 467 AU or almost 12 times as far away as the 40AU orbit of Pluto. According to Sitchin's reckoning, Nibiru's orbit is tipped to the south about 30 degrees to the ecliptic and it orbits retrograde. Orbital mechanics reveals that it would spend the vast majority of its time well beyond Pluto and thus out of sight and a great distance from our sun. Calculations show it would be inside of Jupiter's orbit less than four years.

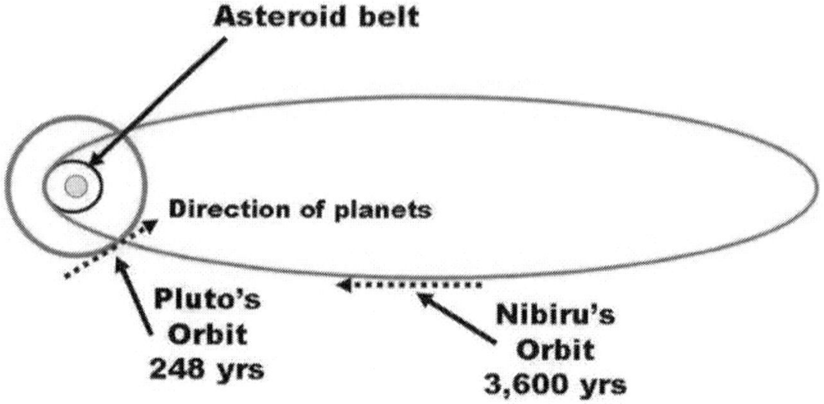

Author's depiction of Nibiru's orbit compared to Pluto and the asteroid belt. Note retrograde motion of Nibiru.

The figure below shows approximately how long it would take Nibiru to transit from X1 where it crosses Jupiter's orbit coming inbound to X2 where it crosses Jupiter's orbit as Nibiru leaves the inner solar system. This probably represents the time it would be visible from Earth by the Sumerians. This graph is approximately to scale so an estimate of the transit time can be made by comparing area swept out as Nibiru sweeps past the Sun. One of Kepler's laws says that as a planet orbits in an ellipse, it sweeps out equal areas per unit of time. The area, A, inside the asteroid belt represents about 24.1 square unites and represents about 4.6 years of transit time. By adding A2 + A1 + a2 we get 18.6 square unites so the transit time from X1 to X2 is 4.6 (18.6 / 24.1) = 3.55 years.

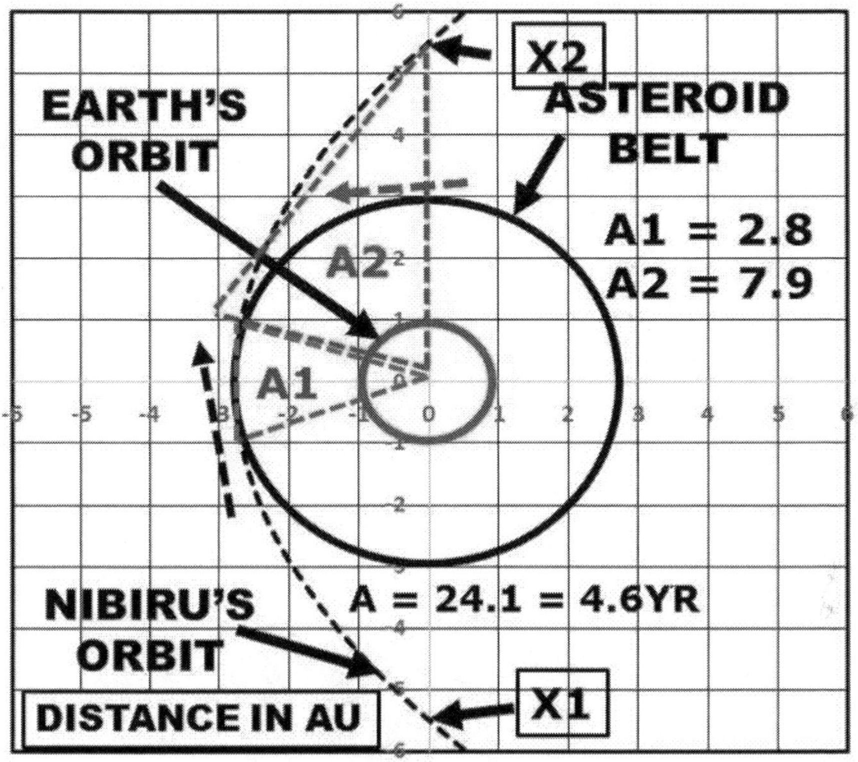

As Sitchin was writing ***The 12th Planet***, there were some astronomers who believed there was a Planet X out there that was perturbing the orbits of the outer planets. One such astronomer was the late Dr. Robert S. Harrington, astronomer at the U. S. Naval Observatory. He was well respected for his work in orbital mechanics. On August 30, 1990, he met with Zecharia Sitchin. It seemed that both men had similar ideas about the location of Planet X. Because they believed the planet would approach from the south, Dr. Harrington went to New Zealand to make observations. In his paper, *Search for Planet X, Reports of Planetary Astronomy, 1991* (see N92-12792 03-89) published in October, he indicated his plans to make photographs in April and June, 1992, with the twin 20 cm astrograph in New Zealand on two successive nights near the time the area of interest was to be in opposition. These plates would be sent back to Washington to be blinked to identify anything that has moved.

Blinking is a process where two photographic plates taken at different times of the same region of stars are compared as though overlaid to detect a change in position of the body of interest. Some believe that he had found Planet X, but he died on January 23, 1993, from cancer of the esophagus, before publishing his results.

Dr. Robert S. Harrington
Credit: © Z SITCHIN www.sitchin.com

At the time ***The 12th Planet*** was published, Pluto was considered a planet. Sitchin made the radicle statement that the Sumerians believed it was a moon of Saturn cast off when

Nibiru made a close encounter on one of its transits around the sun.

Clyde Tombaugh, while searching for Planet X, discovered Pluto at the Lowell Observatory in 1930. Its orbit is highly unusual in that it is tipped 17 degrees from the ecliptic and passes inside the orbit of Neptune. It would seem reasonable to expect that a moon orbiting Saturn in a plane coincident with the ecliptic might be thrown into an odd orbit around the sun with a vector between the ecliptic and Nibiru's orbital plane. Its unusual orbit is making astronomers now wonder if Pluto is a cast-off moon of one of the outer planets. Seventy-six years

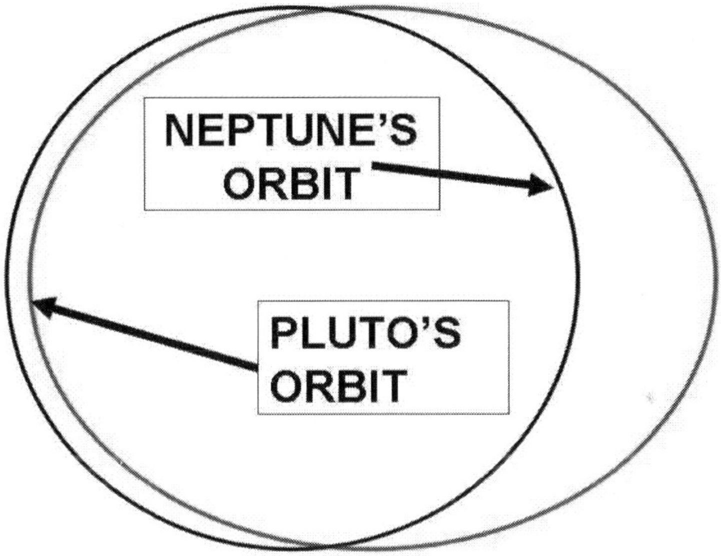

Author's depiction comparing Neptune's orbit with Pluto's.

after Tombaugh's discovery, astronomers have stripped Pluto of its status as a planet. It seems it is too small, and in 2006 the International Astronomical Union reclassified it as a member of the new "dwarf planet" category.

It is now considered a Kuiper belt object. The Kuiper belt was theorized to exist for many years but was not discovered until 1992. It extends from the orbit of Neptune (30 AU) out to about 50 AU. Over a thousand objects large enough to see have been found in the Kuiper belt and it is thought there are over 100,000 thousand smaller objects. Most of the models that are used to predict how the solar system formed require that the total mass of the objects in the Kuiper belt be about 30 Earth masses. In reality the estimated total mass is far less at between five and ten percent of one Earth mass. If the density of the Kuiper belt was always as low as is observed then there would not have been enough material for observed objects to accrete. Also, the inclination of the orbits to the ecliptic and their eccentricity points to some as yet unexplained mechanism.

In November of 2014, astronomers announced some new thinking in regards to our early solar system. They are wondering if a distant planet may lurk far beyond Neptune. It seems that the enormous, stretched orbits of Sedna and 2012 VP113 are so unique that they are unlike anything else in the solar system. Astronomers are now considering the possibility that four billion years ago, as the planets were shuffling for position, something must have "dragged" Sedna and 2012 VP113 from their original smaller orbits into their present orbits. Astronomers are puzzled as to why all the objects beyond 150 AU come closest to the sun at nearly the same time that they cross the plane of the solar system. They are wondering if an unseen planet, a Planet X, in an extremely elongated orbit could be holding the orbits of these and other far-out bodies in their place. In 2014 they speculated that if Planet X exists, it may be anywhere from 250 to 1,000 AU, and in fact a planet twice as massive as Earth must be at least 500 AU from the sun; otherwise, we would have detected it. Some astronomers raise the question: if a remote Planet X exists, how could it form that far from the sun since it would not spend enough time in the planetary "cloud" of planetesimals to accrete into a planet? The most likely answer, they say, is that it is a captured rogue planet. I wonder if these astronomers have read Sitchin's book.

In January, 2016, some astronomers who were perplexed by the fact that at least a dozen Kuiper belt objects have very unusual orbits, announced the possibility of a Planet Nine (IX). The figure below shows a few of the objects of major concern. In each box is the name of the object, its mass relative to Earth, and the angle its orbit makes with the ecliptic.

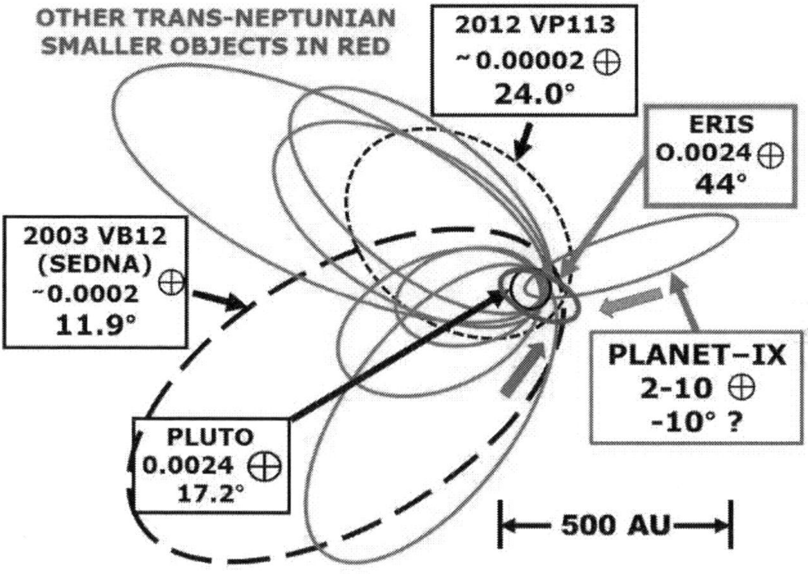

Author's depiction to scale of eight of the Kuiper belt objects with unusual orbits

Pluto and Eris are two of the largest objects found in the Kuiper belt at about 0.0024 times Earth's mass. Sedna and 2012 VP113 are smaller than Pluto and Eris but still larger than those objects shown in red. They all have an unusual behavior in that they all have elongated orbits and they all make their closest approach to the sun as they cross the mid-plane of the solar system at about the same spot. It was estimated that the probability of this happening by chance is 0.007%.

More and more astronomers are beginning to believe these objects were "dragged" out of their normal orbits by some object, Planet IX, that may still be orbiting the sun anywhere from 400 to 1,500 times as far from the sun as Earth. The estimate is that the mass of Planet IX is between two and ten times that of the Earth. The orbit shown in blue is this author's proposal for Planet IX based on the orbit of Nibiru as described by Sitchin. Based on the angle at which the objects were scattered from the ecliptic, this author shows Nibiru coming in from south of the ecliptic at around ten degrees instead of the 30 degrees estimated by Sitchin. Nibiru's orbit is shown going in a retrograde direction with a major axis of about 470.

Another example of the validity of Sitchin's work is his citing Sumerian knowledge of "twin" planets, Uranus and Neptune, which cannot be seen without a large telescope. These two planets are indeed twins with their diameters being the same within about 3 percent. According to the Sumerians, during its orbital passes around the sun, Nibiru would periodically come close to Uranus. During one orbit it came so close it caused it to tip 90° on its axis. Uranus was not known by modern astronomers until William Herschel discovered it in 1781. It was not until 1846 that we discovered Neptune. In 1986, ten years after Sitchin's book, our Voyager fly-by confirmed that Uranus was indeed tipped on its axis by 90 degrees.

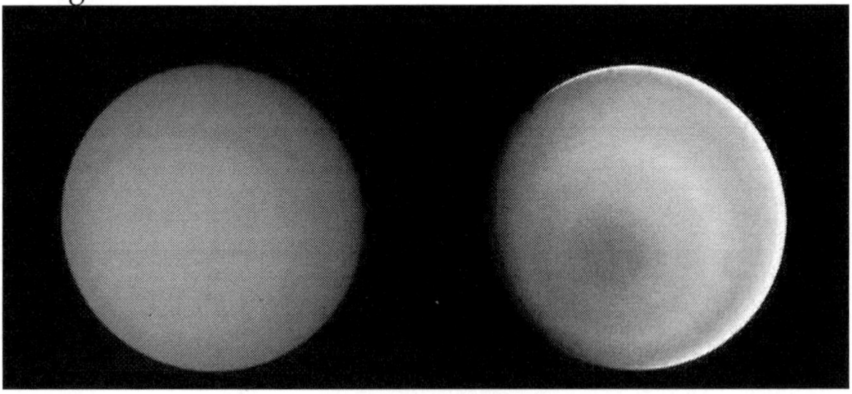

Two images of Uranus as seen in 1986 by Voyager2. Axis Tilt: 97.9° Image credit: NASA/JPL

The similarities between the two planets are striking. Uranus is 4.01 times the diameter of Earth and Neptune is 3.88 times the diameter of Earth.

Neptune as seen in 1989 by Voyager 2.
Axis Tilt: 29.6° Image credit: NASA/JPL

According to Sitchin, Nibiru's orbit takes it as close to the sun as the present asteroid belt. The Sumerians believed that in our early solar system 4.5 billion years ago there was a planet located at that distance that was about twice the size of Earth. They called it Tiamat. It had a host of small moons plus one large one. The Sumerians believed that during one of Nibiru's orbits it encountered Tiamat. Some of Nibiru's moons smashed into Tiamat breaking it in half. Some of the debris along with its large moon was thrown into a lower orbit where it reformed into the Earth and the Moon we have today.

It is interesting to note that today the leading theory for how our moon was formed is known as the Giant Impact hypothesis. It is believed that about 4.5 billion years ago there was a collision between the Earth and another body. One version has named the object Theia and estimated its size to be similar to Mars. Other versions of the hypothesis place the size as closer to the Earth's. Below is an artist's depiction of the "big splash."

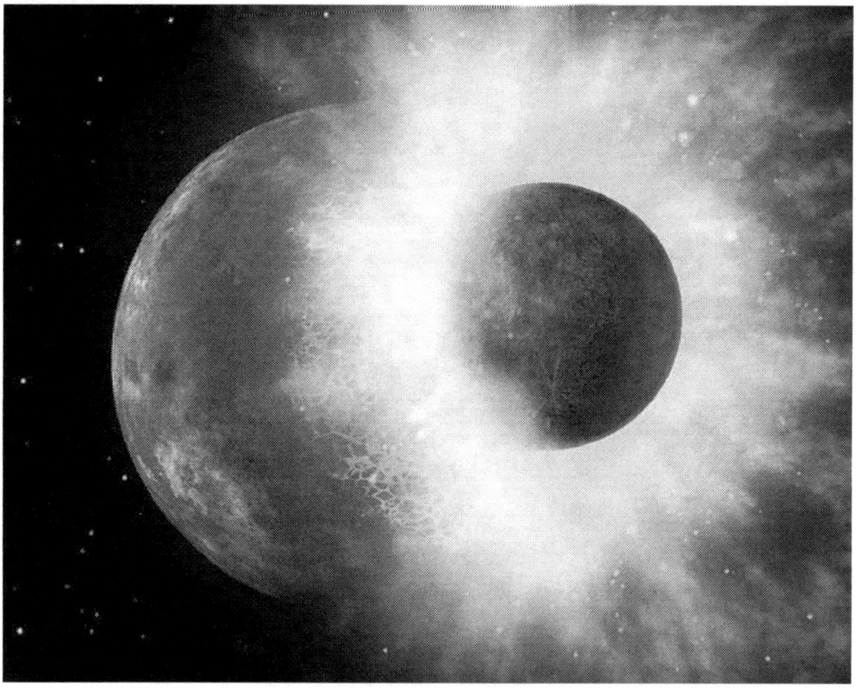

Artist's depiction of a collision between two planetary bodies. Credit: NASA/JPL-Caltech

Much of the debris from the collision between Tiamat and Nibiru's moons remains orbiting as asteroids at between 2.2 and 3.2 AU from the sun. Most asteroids orbit in the same direction as all of the planets but a few dozen were thrown into a retrograde orbit as a result of Nibiru's own retrograde motion.

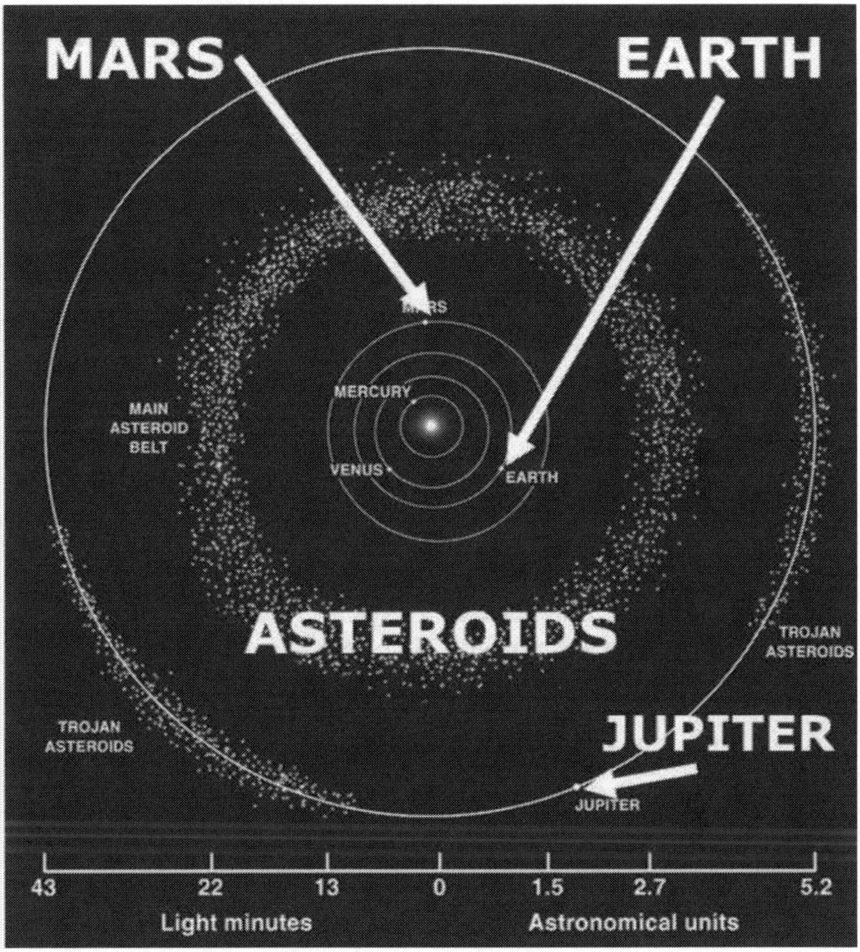

The asteroid belt (white donut-shaped cloud) shown in author's adaptation from a NASA drawing.

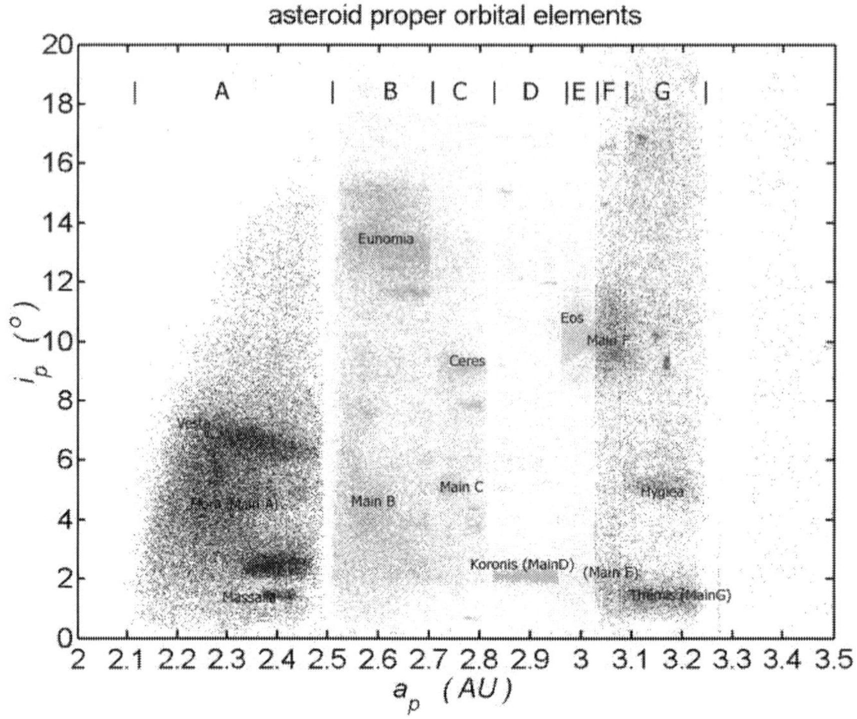

Plot of proper inclination vs. semi-major axis for numbered asteroids. Asteroid families are visible as distinct clumps. Prominent Kirkwood gaps which developed over billions of years divide the core regions (A, B+C, D, E+F+G). Credit: Dreg743 (Wikimedia Commons)

It should be noted that Mars orbits as far out as 1.67 AU and Jupiter as far in as 5.5 AU. Most asteroids orbit in planes outside the ecliptic. Such a large distribution is not surprising if the planetary collision occurred as Sitchin described. Also, note from the figure above that some of the debris is in an orbital plane tipped as much as 20 degrees but the majority have orbits with a tilt centered around eight degrees. Some data, not shown on this graph, indicates a significant number of asteroids have orbital planes tipped as much as 30 degrees. Sitchin said Nibiru crossed the ecliptic at about 30 degrees. Keep in mind that our solar system was very young then and all of the planets were still hot and plastic with the ability to reform into spherical shapes before cooling after a collision.

If the Earth formed where the asteroid belt is today, it might explain a dilemma astronomers have about Earth. Due to its proximity to the sun, it is classified as a rocky planet and should not have as much water as it has. Astronomers are pursuing the hypothesis that during the formation of the solar system, meteors rich in water rained down on the Earth. Recent studies of the atmosphere of comet 67P/Churyumov-Gerasimenko indicate that its water has three times the deuterium-to-hydrogen ratio as the Earth's. It is originally from the Kuiper Belt which is an area rich in comets that extends, as explained previously, from the orbit of Neptune (30 AU) to beyond Pluto (50 AU). It seems that comets are not the answer.

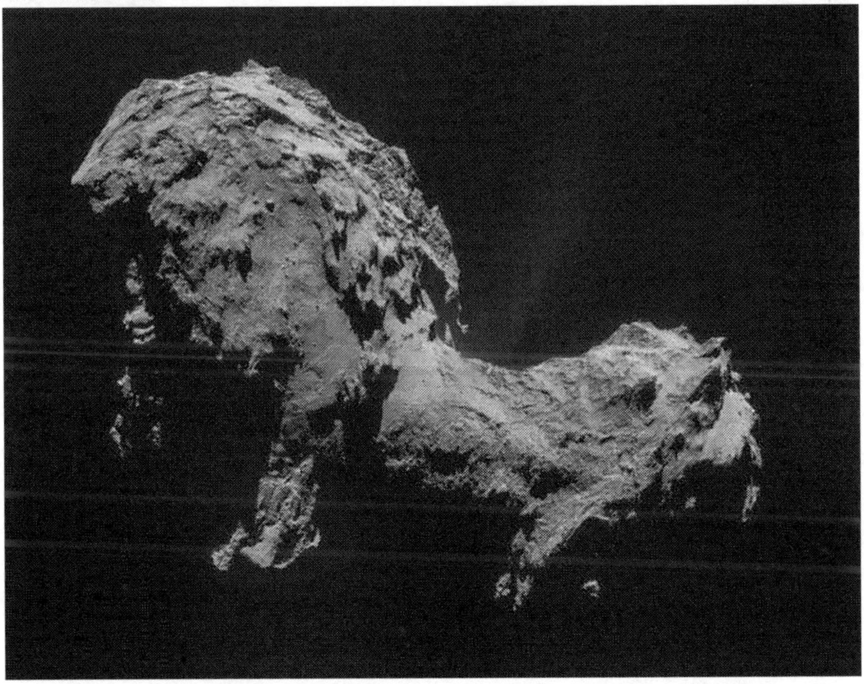

Comet 67P/Churyumov-Gerasimenko as seen by Rosetta.
Credit: European Space Agency

Meteors which come from the asteroid belt are a different story. Chondritic meteorites are rock bearing as opposed to metallic with about 9% known to come from three asteroids. Analysis shows that these have the same ratio of deuterium to hydrogen as is found on Earth.

Eucrite are stony meteorites containing material similar to terrestrial basalts. One was found in northern Africa in 2005 and believed to come from the asteroid Vesta. It also has been found to have the same ratio of deuterium-to-hydrogen as does the Earth. This is a strong clue that the Earth originally came from the asteroid belt region.

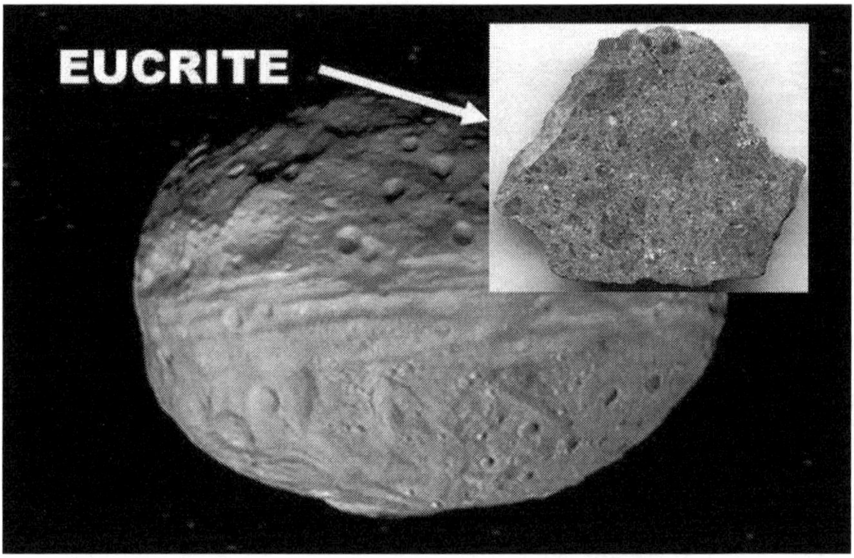

Asteroid Vesta. Credit: NASA/JPL

If the Earth was originally formed in the region of the asteroid belt, as Sitchin says the Sumerians believed, then it would have the same deuterium-to-hydrogen ratio as is seen in the asteroids. Also, it would have as much water as is found in the asteroid belt. Ceres, the largest body in the asteroid belt, has over 25 percent water ice. The farther away from the sun we look, the higher percent of water ice. As we travel farther out to Jupiter and beyond, we find bodies with an even higher percentage of water ice. Ganymede and Europa, two moons of Jupiter, both possess huge amounts of water ice. Ganymede is the largest moon in the solar system and is believed to be equal parts silicate rock and water ice. It is now believed there is a global sea on Saturn's moon, Enceladus.

Ceres, the largest asteroid at 75 miles in diameter, is 25 percent water ice. Credit: NASA/JPL

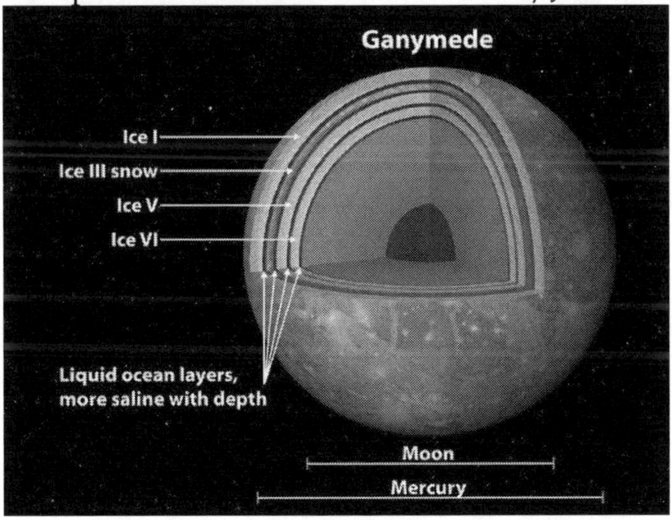

Ganymede with its layers of water ice.
Credit: NASA/JPL-Caltech

Water geysers from an underground sea on Enceladus
CREDIT: JPL-CALTECH/NASA, SPACE SCIENCE INSTITUTE

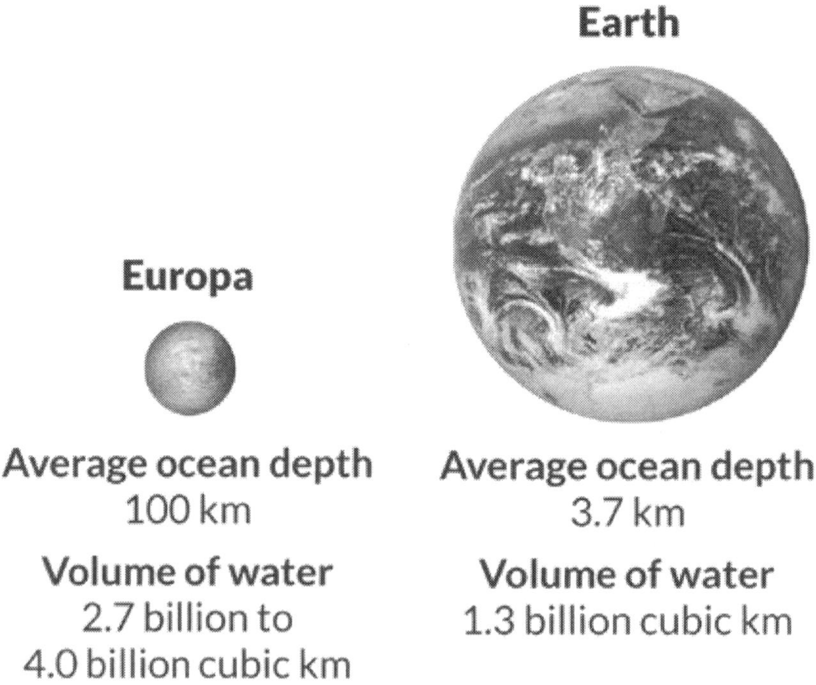

A comparison of Europa with Earth.
Credit: NASA/JPL

ANTHROPOLOGY

According to Sitchin, the Sumerians not only had an understanding of our solar system but also of the origin of humanity. They believed Homo sapiens were created by a genetic manipulation of Homo erectus by beings they called the Anunnaki. Scientists and academics translate Anunnaki as "offspring of Anu." Sitchin's translation is "those who from the heavens came to earth." He believed these were extraterrestrials who came to Earth from Nibiru (or one of its moons) about 445,000 years ago to mine minerals, particularly gold. They had several bases where they did their mining. One was in South Africa. It was there, 300,000 years ago, that they conceived the idea to modify Homo erectus to give them more intelligence and the ability to speak. This would make them more usable as slaves for the mining operations. At first, Homo sapiens could not reproduce, but that was remedied later. They were also given the ability to live long lives with a much slower rate of aging. This would later create a problem of overpopulation.

Clay tablet commemorating Adamu (the first human).
Figure 153, page 352, *The 12th Planet,* Avon Books, July,1978.
Credit: © Z SITCHIN www.sitchin.com

When Sitchin wrote in 1976, genetics was still a new field of study. The human genome had not been completely described. What he wrote then seemed far-fetched. However, subsequent discoveries have given credence to his work. There has never been a discovery of any transitional beings between Homo erectus and Homo sapiens. Ten years after Sitchin published ***The 12th Planet***, Dr. Allan Wilson and two of his Ph.D. students from the University of California at Berkeley, Rebecca Cann and Mark Stoneking, published an article in the January 1986 issue of ***Nature*** that reinforced what Sitchin was saying. The researchers had collected placenta from 147 women around the world and, studying the mitochondrial DNA (mtDNA), they determined that the first Homo sapiens appeared in Africa about 200,000 years ago. This was an order of magnitude fewer years than traditional anthropologists had been using.

The mitochondria are small bodies outside of the nucleus of animal cells. Their function is to produce energy for the cell. In the nucleus, half the genetic material is from the male and half from the female of the species. However, in the mitochondria, the much simpler DNA is contributed almost exclusively by the female. Thus, in the report, the researchers referred to the first human as Eve.

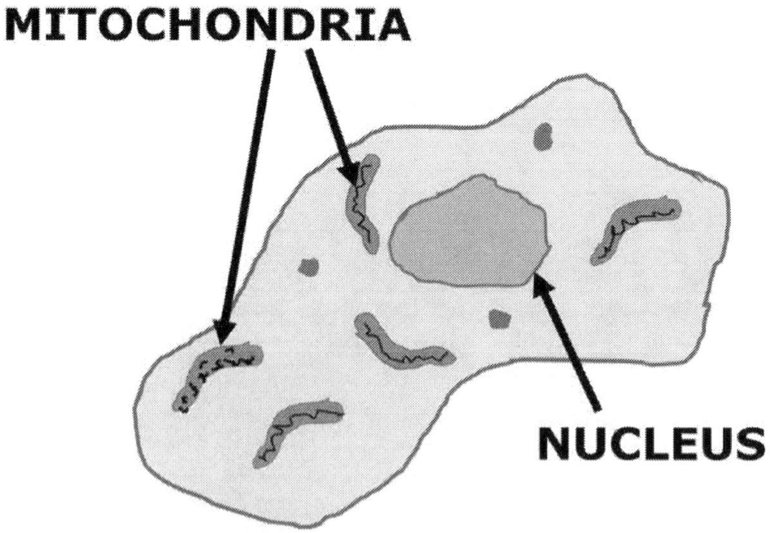

Author's depiction of an animal cell showing mitochondria.

Some people take exception to the findings of Wilson and his team. They believe humans did appear in Africa but only about 5,000 years ago, according to the Bible. They maintain that the scientists made a mistake in determining the rate of mutation of the mtDNA. However, in 2003 an article appeared in ***Nature*** reporting that a skull fragment was found in Ethiopia which was determined to be Homo sapiens and dated to 160,000 years ago. A collection of bones identified as Homo sapiens dated to be about 195,000 old have been found in southwestern Ethiopia at a site called Omo Kibish. These bones include two partial skulls as well as arm and leg bones. Compared to Homo erectus, Homo sapiens have a larger brain and a lower larynx and hyoid bone, making speech possible.

Recently, an ancient megalopolis about the size of Manhattan was found in southern Africa and estimated to be between 160,000 and 200,000 years old. Dan Eden from viewzone.com reports that researcher and author Michael Tellinger and pilot Johan Hine have made the discovery of an ancient metropolis which they found in Africa about 150 miles west of the port of Maputo. Michael Tellinger says that thousands of ancient gold mines have been found in this area over the past 500 years.

Stones arranged into rings can clearly be seen from the air. There is some dispute as to whether these rings are ancient or are simply cattle corrals made by nomadic tribes during the thirteenth century. It is generally believed that domestication of animals did not occur until about 12,000 years ago so if these ruins are nearly 200,000 years old, anthropologists will have to rethink history. (Visit www.viewzone2.com/adamscalendarx.html.)

The sites can be viewed using Google Earth at the following coordinates:

Carolina:	25 55′ 53.28″ S	30 16′ 13.13″ E
Badplaas:	25 47′ 33.45″ S	30 40′ 38.76″ E
Waterval:	25 38′ 07.82″ S	30 21′ 18.79″ E

Machadodorp: 25 39′ 22.42″ S 30 17′ 03.25″ E

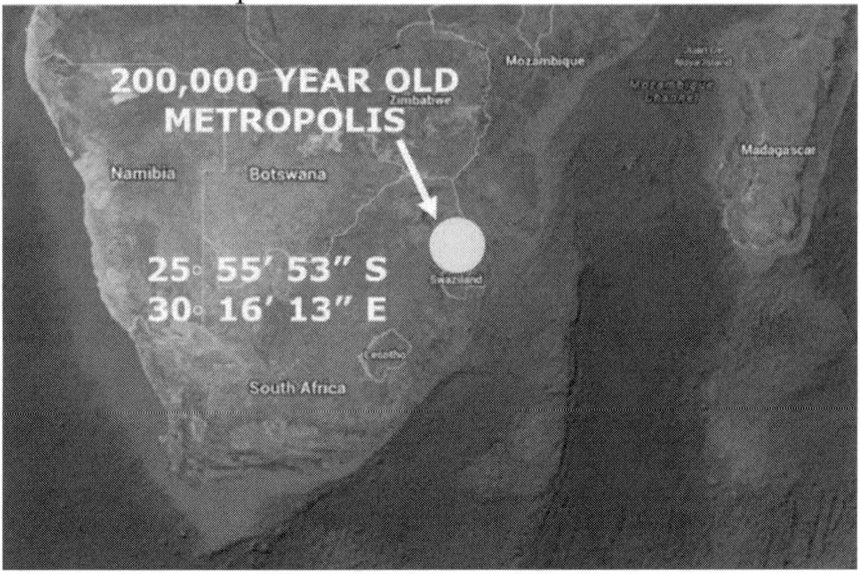

Location of ancient megalopolis.
Author's annotation of source map by Google Maps

Aerial view of stone walls. Credit: Dan Eden, viewzone.com

Carbon dating of the site was not possible because no artifacts were found that were suitable for such dating methods. Recognizing that the stones were aligned at cardinal points, an investigation was made to compare the rising of Orion's belt when it was level with the horizon and with the stones. The most recent estimate based on archaeoastronomy and the erosion of dolerite stones found at the site indicate an age of at least 160,000 years.

Alignment of stones at cardinal points. Credit: Michael Tellinger, viewzone.com

It is thought that mining technology in southern Africa may go back over 100,000 years. The largest gold producing area on Earth is in this same region of southern Africa. According to legends of Zulu medicine man Vusamazulu Credo Mutwa in his book ***Indaba, My Children***, the mines were worked by "artificially produced flesh and blood slaves created by the First People." In addition to gold, that part of the world is rich in other minerals such as chromite, cobalt, manganese, phosphate rock, natural diamonds, and uranium. All important in today's economy.

FLOOD STORY

Sitchin also explored the genesis of Noah's story in ***The 12th Planet***. He was intrigued by the theory of glaciologists, Dr. John T. Hollin, Ph.D., is a Fellow Emeritus of the Institute of Arctic and Alpine Research (INSTAAR) at the University of Colorado. In 1972 he made reference to glaciologist Professor Allan T. Wilson's work at Victoria University in New Zealand. Hollin stated, "Wilson's theory of ice ages implies that the present interglacial will end with, or at least be interrupted by, an Antarctic ice sheet 'surge.' Such surges in the past would have caused distinctive rises of sea level: by 10–30 m, in 100 years **or much less** [bold added], and precisely at the break of climate at the end of each interglacial."

Sitchin felt this might have been the cause of Noah's Flood. He placed the date 13,000 years ago because of the ending of the ice age combined with the return of Nibiru to the region of the asteroid belt. He believed that the unstable ice caps in Antarctica broke loose due to the gravitational influence of Nibiru, similar to the way the Moon causes tides. The discussion given in Appendix B explains how Nibiru's gravitational influence was probably not the cause of the breakaway. This is assuming Nibiru's closest approach to the sun is the asteroid belt. The most likely cause was the effect of moulins and disintegration of ice sheets during the Bolling/Allerod interstadial warming period about 14,700 years ago. Despite the cause of the Deluge, Sumerian history is divided into two periods: pre-deluge and post-deluge.

KINGS LIST BEFORE THE DELUGE

One of the ancient tablets translated by Sitchin and others is known as the Kings List Before the Deluge. Their ruling period is expressed in shars. It is believed that one shar is equal to 3,600 of our years. Sitchin believed a shar was based on the orbital period of Nibiru. According to his translations and those of others, before the Deluge, eight kings ruled for over 69 shars with the longest reigning for 12 shars. That means that one of the kings ruled for 43,000 years! How is this possible? Perhaps it is possible because, according to Sitchin, these rulers were not human but were aliens, the Anunnaki. Who knows how long they live? If they are capable of interplanetary travel, then with their technology, they probably live as long as they want. Consider that fruit flies live only 15 days and humans live about 2,000 times as long. Fruit flies must think we live forever. If the Anunnaki live 2,000 times as long as present-day humans, they could live to be 160,000 years old! 43,000 years would only represent 25 percent of a typical Anunnaki's life span.

Kings List Before the Deluge, Schøyen Collection MS 2855.
Credit: The Schøyen Collection, Oslo and London

KINGS LIST AFTER THE DELUGE

It would have taken hundreds or perhaps thousands of years for the lower region of Mesopotamia to recover from the Deluge (or series of deluges). Oral tradition of the flood story would probably have made humanity slow to return. The devastation in Japan from the 2011 tsunami pales in comparison to the devastation that probably occurred in southern Mesopotamia.

Kings List After the Deluge, Schøyen Collection MS 1686
Credit: The Schøyen Collection, Oslo and London

Another of the ancient tablets is known as the Kings List After the Deluge. According to this tablet, after the Deluge, humans were allowed to rule. The first king ruled 1,200 years. Was this another alien? No, Sitchin said this was a human who was given the right to rule. After the Deluge, the rulers of the ancient Sumerian cities ruled at the pleasure of the gods (lords).

1,200 years of rule by a human might seem hard to believe, but recall from the Old Testament that people around Noah's time lived over 900 years. In fact, Noah lived to be 950. Some have argued that the numbers reported in the Bible represented months not years. Thus, Noah only lived to be 79 if that were true. If indeed it was months that were counted as years, some of the men mentioned in the Book of Genesis were quite young when they fathered children. For instance, Noah's grandson, Arpachshad who lived to be 436, had a son, Shelah, when he was 35. If that were months, then Arpachshad was less than three years old when he fathered children! According to Genesis (King James Version), assuming the years that were reported were actually months, Kenan was 5.9 years old when he fathered Mahalalel (Genesis 5:12) and Mahalalel was 5.4 years old when he fathered Jared (Genesis 5:15). Clearly, Noah really did live 950 years and not 950 months.

The fact that we do not live that long today is a question for geneticists not historians. It is as though there were some mutation occurring in whatever gene controls our lifespan. The cause for aging is under investigation, and there is strong suspicion that the answer lies in genetics. Since 1950, our lifespan has increased by about 5 years every 25 years. With the expected dramatic increase in medical technology, the life expectancy curve will begin to bend upward exponentially. Already, pharmaceutical companies are learning how to tailor medicine based on the patient's genetics. Soon we will be able to build replacement body parts from the patient's own tissue, solving the rejection problem we have today.

Geneticists have already been able to extend the life of their lab animals by up to ten-fold. Genetics researchers are able to study the function of genes by using what they call RNA inhibitors. Using these inhibitors, they are able to turn off selected genes and then study what happens. These researchers have been able to extend the lives of roundworms ten-fold from 14 days to 140 days through genetic manipulation. If they could do that to humans, perhaps we would live to be as old as Methuselah.

Researchers are studying people who live past 100 years to find out why they live so long. They have discovered that there is a gene which they referred to as FOXO3A which may be a global factor in longevity. Everyone has this gene, but variations appear to give centenarians the ability to live past a century. By manipulating this gene, they have extended the life of some mice. Researchers hope that by understanding how variance of FOXO3A changes some proteins, perhaps they could produce drugs that replicate that change and give our bodies longer life. It would not be surprising to learn that the Anunnaki already knew how to do this.

We now turn to examining some of the ancient flood stories from Mesopotamia as well as the story of Noah from Genesis. Any hypothesis about the cause of the Deluge should be supported by historical records, such as they are.

Chapter 3
Ziusudra and Atra-Hasis

This chapter is important because the genesis of Noah's story can be found in ancient stories from a region of Mesopotamia where the Sumerian civilization existed 6,000 years ago. The Atra-Hasis as well as the Epic of Gilgamesh were assembled from bits and pieces of tablets found in various regions around Mesopotamia. The translations of these stories available to the public represent the consensus view of experts in the field. Often when tablets are found, they are broken or incomplete. Combine that with the fact that each city often had its own slightly different version of these stories, and one can realize the tremendous effort required to get a complete story. The translations given below represent the consensus views of experts in the field.

ATRA-HASIS

We will begin with the oldest story, Atra-Hasis. One of the best English translations is by W. G. Lambert and A. R. Millard in their book ***Atra-Hasis: The Babylonian Story of the Flood***. The original clay tablets are dated to about 3,200 BC. The old Babylonian version has been dated to about 1,700 BC. Keep in mind that the dating of the tablets does not reflect the date of the flood, only the date it was recorded after humans finally learned to write. The following is a summary of the translation by Lambert and Millard.

Atra-Hasis (also known as Ziusudra) was the king of Shuruppak. This was before the Deluge when the land was ruled by the Anunnaki, the gods (lords). (Note: Nefilim [Nephilim] were mentioned as the sons of God in the Old Testament.) In the Atra-Hasis story, the Anunnaki were concerned because of human overpopulation. Recall that humans back then did not age as fast as we do today so one man and woman could produce hundreds of offspring in their lifetimes and their offspring likewise. Thus, a population explosion could have occurred. Drought and pestilence had not corrected the problem of overpopulation, so the Anunnaki needed another plan. Enlil, the chief god, was aware that a catastrophic deluge was about to happen in Mesopotamia (their technology would have allowed them to know this) and decided not to warn the humans and to let them drown; nature would take its course.

Author's depiction of Atra-Hasis's kingdom before the Deluge. Source map: http://www.mideastnews.com/kuwait26.html

Enki, one of the Anunnaki who had helped create the humans, had a favorite human who was loyal to him. His name was Atra-Hasis (Ziusudra, the wise one). Enki wanted to warn him, but Enlil made all the gods including Enki take a pledge not to do so. Enki cleverly got around his pledge by standing outside the hut of Atra-Hasis and talking to the wall, not directly to Atra-Hasis. Enki spoke to the wall and said, "Reject possessions and save living things." He then went on to detail the construction of an ark saying, "Roof it so the sun cannot shine in" and "make upper and lower decks . . . the bitumen strong to give strength." Enki told Atra-Hasis that he must convince the townspeople to help him construct the ark. The floods came and "roared like a bull . . . darkness was total." The flood lasted seven days and seven nights. After the ark came aground, Atra-Hasis made a burnt offering. "The gods smelt the fragrance and gathered like flies." Realizing their mistake in trying to eliminate the humans, they repented and blamed Enlil for nearly destroying humanity. When Enlil arrived on the scene, he was furious at first but then saw Enki's wisdom. He vowed to find a better way to control the population. It is not clear what happened after that because of the poor condition of the tablets. (Perhaps they genetically altered humans to allow aging?)

Author's depiction showing how the Deluge carries Atra-Hasis north toward Turkey. Source map: http://www.mideastnews.com/kuwait26.html

Below is a listing of lifespan data taken from Genesis. Note that before the flood, people lived 900+ years but after the flood, beginning with Noah's grandson, people lived shorter and shorter lifespans. Was this genetic engineering at work? Not shown in the second figure is Moses who only lived to be 120.

LIFESPAN BEFORE THE FLOOD

PATRIARCH	LIFESPAN (yrs)
ADAM	936
SETH	912
ENOSH	905
KENAN	910
MAHALALELI	895
JARAD	962
ENOCH	365 – GOD TOOK
METHUSELAH	969
LAMECH	777
NOAH	950

Lifespan before the Flood. From Genesis chapter 5 (KJV)

LIFESPAN AFTER THE FLOOD

PATRIARCH	LIFESPAN (yrs)
ARPACHSHAD (from Shem)	438
SHELAH	433
EBER	464
PELEG	239
REU	239
SERUG	230
NAHOR	148
TERAH	205
ABRAHAM	175

Lifespan after the Flood. From Genesis chapter 11 (KJV)

Chapter 4
Epic of Gilgamesh

The Flood Tablet, relating part of the Epic of Gilgamesh, from Nineveh, northern Iraq, Neo-Assyrian, 7th century BC © Trustees of the British Museum. Photo ID: 00396940001

The following description is taken from the English translation by Stephanie Dalley. Comments within square brackets [ ] are mine.

In the Gilgamesh epic, Gilgamesh was king of the city of Uruk and the supreme hero of this Sumerian myth. There are about twelve tablets, but only the eleventh deals with the flood story. In the eleventh tablet Gilgamesh went on a quest to find eternal life which he had heard was achieved by a man named Utnapishtim. Below is a map of the region after the Deluge which shows the location of Gilgamesh's kingdom and how it relates to Shuruppak; Kish, the first city after the flood; and Ur, Abraham's birthplace (according to Sir Leonard Woolley).

Author's depiction of Gilgamesh's kingdom as it relates to Abraham's birthplace in Ur (according to Woolley).
Source map:
http://www.mideastnews.com/kuwait26.html

In tablet 11 of the story, Gilgamesh found Utnapishtim near the headwaters of the Euphrates River where the tall cedars grow [Turkey]. Utnapishtim, who had been king of Shuruppak, tells Gilgamesh the story of the Deluge. The following are excerpts from Dalley's translation.

Utnapishtim explains, "Shuruppak . . . was already old when the gods within it decided [to] make a flood." [Note that in this version of the Deluge it was the gods who **caused** the flood.]

Below is this author's drawing showing Utnapishtim's kingdom. Note the marshlands on the north end of the lake that would later become the Persian Gulf.

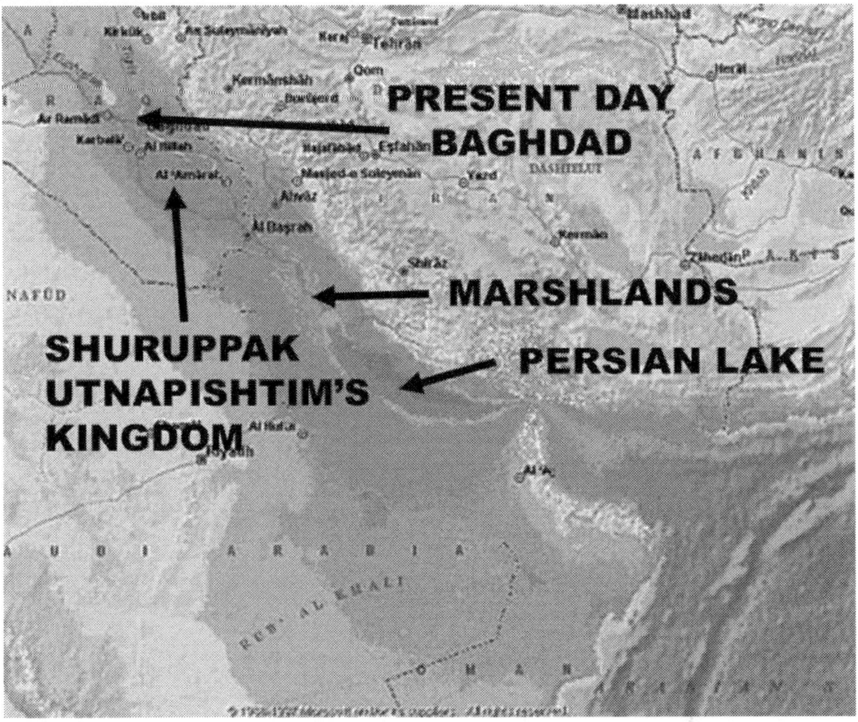

Author's depiction of Utnapishtim's kingdom before the Deluge. Source map: http://www.mideastnews.com/kuwait26.html

Below is some dialog given in Stephanie Dalley's book. I have paraphrased and shortened some of her translations. Comments within square brackets [ ] are mine.

"Ea [Enki] took an oath with them [the other gods], so he repeated their speech to a reed hut" [Utnapishtim's hut].

"Man of Shuruppak . . . dismantle your house and build a boat."

"Leave your possessions . . . put aboard the seed of all living things . . ."

"The boat . . . shall have her dimensions in proportion, her width and length shall be in harmony."

"My master . . . how can I explain myself to the city?"

"You shall speak to them thus: 'I think that Ellil [Enlil] has rejected me, and so I cannot stay in your city . . . I must go to the Apsu [this was the swampy area north of the Persian "lake"] and stay with my master Ea [Enki]. Then he will shower abundance upon you.'"

[Enki gave details on the construction of the ark. Utnapishtim enlisted the help of the townspeople to build an ark. Within seven days the ark was completed and launched into the Euphrates River. Then Utnapishtim said,]

"I loaded her with everything there was . . ."

"The storm was terrifying to see."

"I went aboard the boat and closed the door."

"A black cloud came up from the base of the sky. Adad [Storm god] kept rumbling inside it."

"The Anunnaki had to carry torches, they lit up the land with their brightness."

"On the first day the tempest (rose up), blew swiftly and (brought) the flood weapon."

Author's depiction showing Deluge carrying Utnapishtim north toward Turkey. Source map: http://www.mideastnews.com/kuwait26.html

"Even the gods were afraid . . . they went up to the heaven of Anu [King of the gods]."

"For six days and seven nights the wind blew, flood and tempest overwhelmed the land."

"The sea became calm . . . I looked at the weather: silence reigned . . . The floodplain was flat as a roof."

[He was drifting northward, far from land on either side but because of the Earth's curvature could see no land. He probably thought the whole world was covered with water.]

"The boat had come to rest on Mount Nimush . . . (even until) Nimush held the boat fast."

"When the seventh day arrived, I put out and released a dove . . . it came back . . . a swallow . . . it came back . . . a raven . . . it did not come back."

[Utnapishtim then unloaded the boat and arranged seven jars filled with incense.]

"The gods smelt the fragrance . . . and like flies, gathered over the sacrifice."

"As soon as the Mistress of the Gods arrived she said . . . 'Let other gods come . . . but let Ellil not come . . . because he did not consult before imposing the flood.'"

[As soon as Ellil arrived, he was furious that humans had survived but was soon convinced that it was for the best.]

"Ellil came up into the boat and seized my hand and led me up . . . and led my woman and made her kneel at my side."

"He touched our foreheads . . . and blessed us: 'Until now Utnapishtim was mortal, but henceforth . . . shall be as we gods are. Utnapishtim shall dwell far off at the mouth of the rivers.'"

[These were the Tigris and Euphrates. Note: The gods (lords) were immortal in the eyes of the ancients.]

Chapter 5
Noah's Flood in Genesis

Below are verses from the King James Version of the Bible. My comments are in square brackets [ ].

From Genesis 6:

- 13 And God said to Noah, "I have determined to make an end of all flesh, for the earth is filled with violence because of them; now I am going to destroy them along with the earth."
- 14 Make yourself an ark of gopher wood [cypress]; make rooms in the ark, and cover it inside and out with pitch.
- 15 This is how you are to make it: the length of the ark three hundred cubits [450 ft], its width fifty cubits [75 ft], and its height thirty cubits [45 ft].
- 16 Put the door of the ark in its side; make it with lower, second, and third decks.
- 17 For my part, I am going to bring a flood of waters on the earth, to destroy from under heaven all flesh in which is the breath of life; everything that is on the earth shall die.
- 18 But I will establish my covenant with you; and you shall come into the ark, you, your sons, your wife, and your sons' wives with you.
- 19 And of every living thing, of all flesh, you shall bring two of every kind into the ark, to keep them alive with you; they shall be male and female.

From Genesis 7:

- 4 For in seven days I will send rain on the earth for forty days and forty nights; and every living thing that I have made I will blot out from the face of the ground.

- 11 In the second month, on the seventeenth day of the month, . . . all the fountains of the great deep burst forth, and the windows of the heavens were opened.
- 12 The rain fell on the earth for forty days and forty nights.
- 24 And the waters swelled on the earth for one hundred and fifty days [covering the mountain tops by 20 feet].

From Genesis 8:

- 3 At the end of one hundred and fifty days the waters had abated.
- 4 In the seventh month, on the seventeenth day of the month, the ark came to rest on the mountains of Ararat.
- 10 He waited another seven days, and again he sent out the dove from the ark.
- 11 And the dove came back . . . with an olive leaf.
- 12 Then he waited another seven days, and sent out the dove; and it did not return.
- 14 In the second month, on the twenty-seventh day of the month, the earth was dry.
- 15 Then God said to Noah, 16 "Go out of the ark."
- 20 Then Noah built an altar to the Lord, and . . . offered burnt-offerings on the altar.
- 21 And when the Lord smelt the pleasing odour, the Lord said in his heart, "I will never again curse the ground because of humankind . . . nor will I ever again destroy every living creature as I have done."

[Noah lived another 350 years after the Flood.]

COMPARING AND CONTRASTING THE THREE STORIES

It is interesting to compare and contrast these three versions to see how the story evolves. In Atra-Hasis **the gods knew** the flood was coming but chose not to warn humanity as a way to control the population growth. In the Epic of Gilgamesh, **the gods caused** the flood to eliminate the humanity they had begun to dislike. In Noah's Flood from Genesis, **God [singular] caused** the Flood to eliminate all life which he deemed evil. In all three versions, the hero releases birds to determine when to disembark. In all three versions the hero makes an offering after surviving which pleases the gods (God) who then repent.

Now we will consider some hypotheses about the cause of the Deluge which became known as Noah's Flood.

Chapter 6
Was a pole shift to blame?

Is it possible that Noah's Flood was caused by a shifting of the Earth's poles resulting in a sinking of land and rising of seas? Author John White has compiled numerous theories about major pole shifts that may have occurred on Earth. Many of the theories put forth offer logical and convincing arguments for the probability that the Earth may have had a significant shift of its poles.

One of the arguments for a pole shift is that plants and animals are "out of place" considering their normal environment. For example, how does one explain woolly mammoths found in Siberia? Also, there is evidence written in legends found in the history of the Greeks, Egyptians, and Hopi, as well as the Eskimos. Some of these ancient legends talk about a sudden shifting of the stars or a change in the location of the sun or Moon. When missionaries encountered the Eskimos in Greenland, they were told that in ancient times the Earth turned over. All of these would indicate a sudden pole shift.

There is strong evidence that the magnetic poles have shifted and continue to shift from time to time. This does not mean that the axis of rotation shifts. The Earth's magnetic field is caused by electric currents in its molten iron core. These currents drift by as much as 10 miles per year. Anyone who navigates by compass must account for these magnetic variations. There is evidence that every half million years or so the magnetic poles completely flip so the north magnetic pole becomes south. This was determined by studying the alignment of the magnetic fields frozen in lava that seeps out of the ridges in the ocean floor. It is unlikely that shifting magnetic

poles, even if sudden, would result in the deluge Noah actually
experienced.

For our discussion we must keep in mind that Noah's Flood was a very short-term event relative to the history of the planet. If the cause was a pole shift, it had to be a sudden one that involved the shifting of the Earth's axis of rotation, or the surface relative to the axis, in only a few hours or days. Two of the theories discussed in John White's book involve just such an event.

Hugh Auchincloss Brown was an electrical engineer who had an interest in gyros. He became interested in stories about mammoths found frozen in the Arctic with fresh food in their mouths. To him, this indicated a sudden death by freezing. Combining this information with his knowledge of gyros brought him to the conclusion that the Earth may have flipped on its axis. This, he reasoned, was caused by a torque produced by the huge amount of ice built up off-center from the North and South Poles' axis. He felt that the sudden shifting of the poles would cause earthquakes, massive floods, and other devastation. His calculations showed that this flipping occurred about every 8,000 years and was imminent. In 1967 he published his book ***Cataclysms of the Earth***. From 1948 until his death in 1975, he prophesied a doomsday caused by the shifting of the poles. Concern for this imminent disaster prompted him to call for the use of nuclear explosions to break up the ice at the South Pole.

Charles Hapgood, a history professor at Springfield College in Massachusetts and later Keen State University in New Hampshire, came to a similar conclusion but with a slight variation. Hapgood's view differed from Brown's in that he did not visualize the Earth's axis as shifting but rather just the outer skin or crust slipping over the mantle. His interest was piqued in 1949 as a result of a class project he gave on the topic of Atlantis. He concluded that the answer to

Atlantis was not to be found in archeology but in geology. His investigation led him to a similar conclusion as that proposed by Brown. In 1958 he published ***Earth's Shifting Crust: A Scientific Key to Many of Earth's Mysteries***. There was considerable credibility to this book as it had a forward by none other than Albert Einstein. In 1970 he published a later version entitled ***The Path of the Pole***.

He agreed with Brown that the driving force would be the huge ice buildup off-center from both the North and South poles. Keep in mind that the Earth's crust varies from 3 to 44 miles which on average is only about 0.5 percent of the Earth's diameter. This represents a very small percentage of the Earth's total moment of inertia. He visualized that the layers beneath the crust were thyrotrophic in the sense that under a static situation they appear rigid but under a sudden stress behave like a plastic or viscous fluid. For example, some clays behave that way and have been the cause of sudden landslides. The behavior of ketchup is another good example. Hapgood reasoned that, with sufficient torque, the crust could slip over the mantle to bring things into balance. This slippage would occur in a very short period of time, perhaps hours. The results would be traumatic earthquakes and floods.

Chapter 7
Noah's Flood per Ryan and Pitman

At the time Ryan and Pitman wrote their book on Noah's Flood, they were senior scientists at the Lamont-Doherty Earth Observatory of Columbia University. Founded in 1949, it has become the leader in earth sciences especially in marine geology and is a member of the Earth Institute which works to solve some of the world's problems. Lamont-Doherty is the world's largest repository of core borings retrieved from the depths of the oceans. They were the first to map the seafloor and develop a computer model that confirmed plate tectonics.

It was during their work in the area of the Black Sea that Ryan and Pitman developed their hypothesis about the genesis of Noah's story. Through core samples and sonar analysis of the floor of the Black Sea and the Bosporus Strait, they discovered that 7,600 years ago there had been a large freshwater lake 300 feet below present sea level. 20,000 years ago, during the maximum of the last ice age, sea levels were about 370 feet below present. Water was locked up in the Laurentide Ice Sheet up to two miles thick in the northern hemisphere and in the Antarctic glaciers up to eight miles thick in the southern hemisphere.

Author's annotation of artist's view of ice age at global maximum. Source: Lttiz, Wikimedia Commons

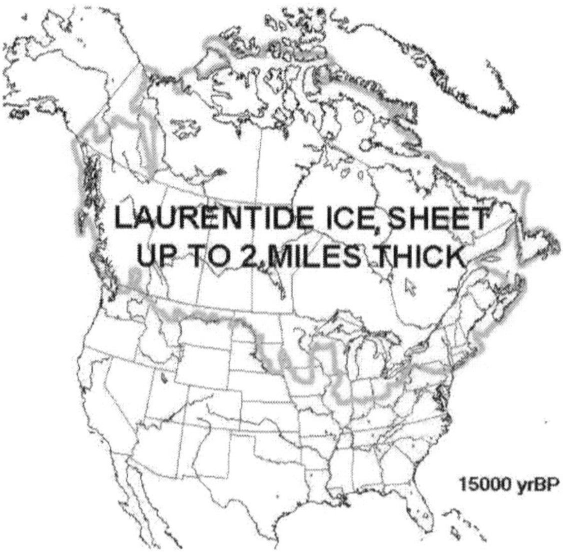

Laurentide Ice Sheet (outlined) 15,000 years ago
Credit: NOAA

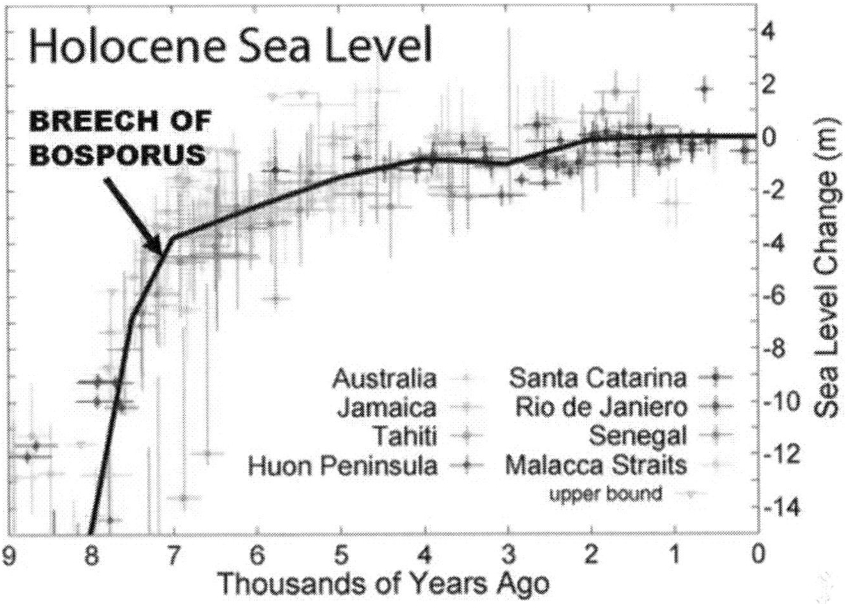

Author's depiction showing when the Bosporus was breeched. Source graph: Robert A. Rohde - Global Warming Art project. (Wikimedia Commons) from Source: Fleming et al. 1998, Fleming 2000, & Milne et al. 2005

Ryan and Pitman hypothesized that there were people living around what they called the New Euxine Lake. 7,600 years ago, as the Earth exited the ice age and the Laurentide Ice Sheet collapsed in North America, the water level in the Aegean Sea rose to the point where it began rushing through the Bosporus Strait. It eroded a channel so large that the water flowed down into the lake at a rate 400 times greater than the flow over Niagara Falls. The water level of the lake rose six inches per day, so fast that very few inhabitants had a chance to escape and many drowned. Those who did escape would have recorded this as an historic event. The authors believe this is the genesis of Noah's Flood story. In the figure below, it is worth noting the proximity of Mount Ararat.

Bosporus Strait today. Source map: Google Maps

Noah's Flood was an interesting book and got me thinking about whether such an event may have occurred in Mesopotamia. So, I began studying pole shift theories, recorded changes in sea level, mega-tsunamis, and maps of the Persian Gulf.

Chapter 8
Noah's Flood per Dr. Bruce Masse

Dr. W. Bruce Masse is a scientist at Los Alamos National Labs near Santa Fe, New Mexico. He made the observation that the scientific study of events that occurred in antiquity could be aided by a study of the local culture, traditions, and myths. He thought that by taking myths and legends at face value and using scientific methods, it might be possible to extrapolate into reality.

He became interested in applying this technique to the flood story in the Old Testament and began studying many of the flood myths from around the world. Guessing that Noah's story may have its roots in and around the Indian Ocean, he developed his hypothesis that either an impact event or major landslide in that region resulted in a mega-tsunami which produced the deluge described in the stories.

At a conference of geologists, astronomers, and archeologists in 2004, Masse presented his hypothesis and attracted considerable interest. In 2005, Dr. Masse formed a group of interested parties called the Holocene Impact Working Group to continue the investigations. One member of the group was Dr. Dallas Abbott from Lamont-Doherty Earth Observatory of Columbia University. When she began searching with Google Earth, she saw chevrons along shorelines around Africa, Asia, and southern Madagascar. The ones in Madagascar all pointed in the same direction. From the size and shape, she thought they may have been formed by huge waves from a tsunami triggered by an impact of a comet in the Indian Ocean.

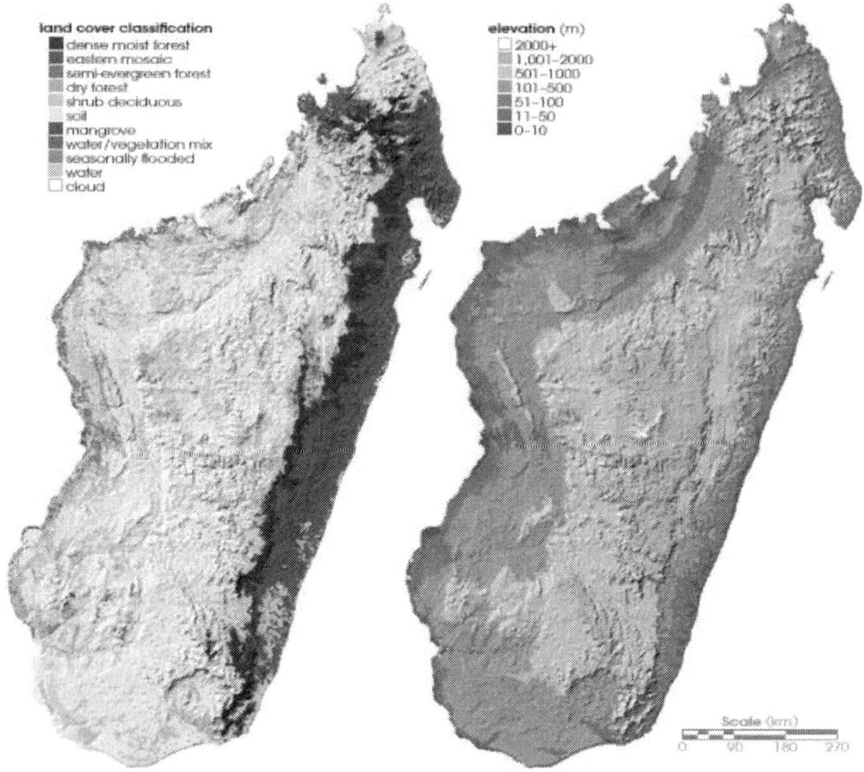

Topographic map of Madagascar. Credit: NASA

An International Tsunami Expedition to southern Madagascar was made between August 28 and September 12 in 2006 to gather firsthand data on these "chevrons." The team investigated four sites and recorded data from mega-tsunami chevrons at Faux Cap, the largest; at Fenambosy and Ampalaza; and in marine sediment dumped at Cape St. Marie. Dr. Abbott reported that the maximum run-up (height above surrounding terrain) was 205 meters (673 feet) with the maximum length being 45 km (28 miles). The largest dune covered an area larger than the island of Manhattan. In addition to sand, the dunes contained tiny fossils from the ocean floor. A close-up inspection revealed iron, nickel, and chrome fused to the fossils. Rocks and sea shells also littered the dunes.

Using Google Earth and following the path indicated by the chevrons, she was led to the discovery of a huge impact crater in the Indian Ocean. It is named Burckle Crater and is located at coordinates 30.865°S and 61.365°E. It is 29 km (18 miles) in diameter and 12,500 feet below the Indian Ocean surface. I was not able to get sufficient resolution on my version of Google Earth to see evidence of an impact crater. The figure below shows a view representing Masse's hypothesis.

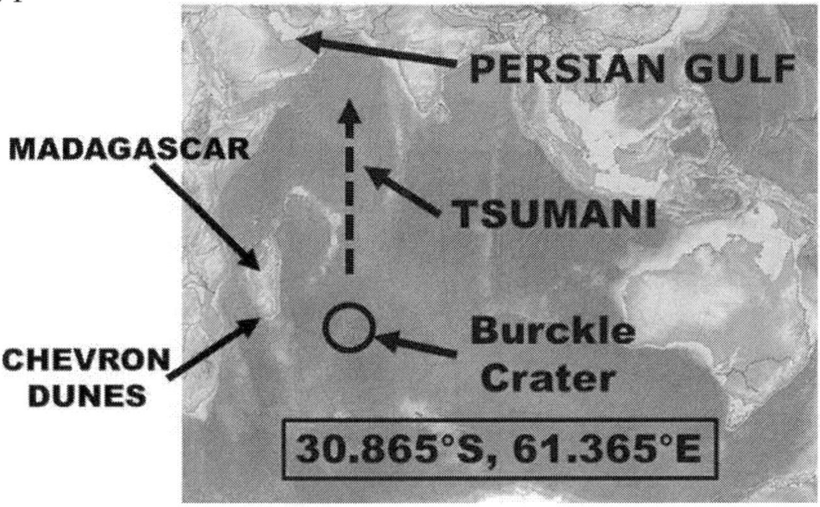

Author's depiction of relationship between Madagascar and Burckle Crater. Source map: Google Maps

Tip of the Fenambosy chevron deposit in southern Madagascar. Credit: Dallas Abbott, Lamont-Doherty Earth Observatory

Masse's hypothesis is that the impact created both the mega-tsunami that caused the chevron dunes seen in southern Madagascar as well as the mega-tsunami that traveled up the Persian Gulf to become Noah's Flood.

Dunes relative to Burckle Crater.
Source map: Google Earth

Note: The vector to crater lines up with the direction of the chevron dunes as shown below.

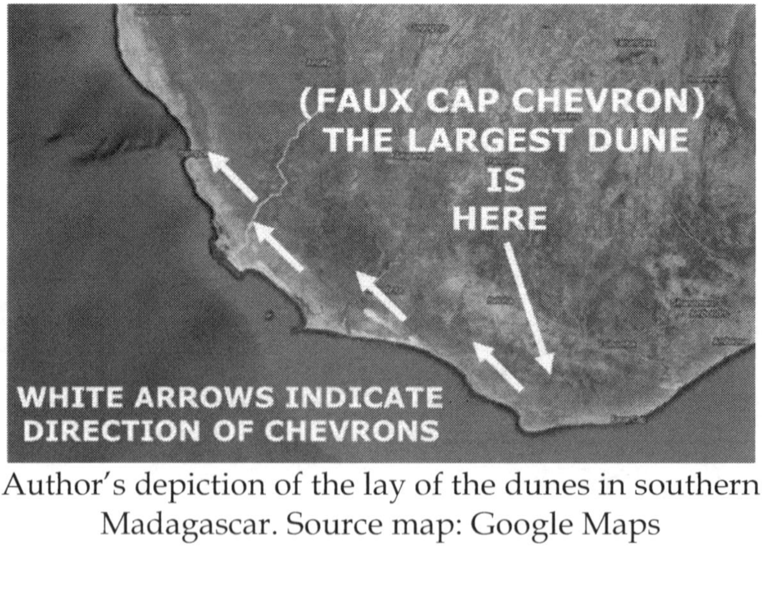

Author's depiction of the lay of the dunes in southern Madagascar. Source map: Google Maps

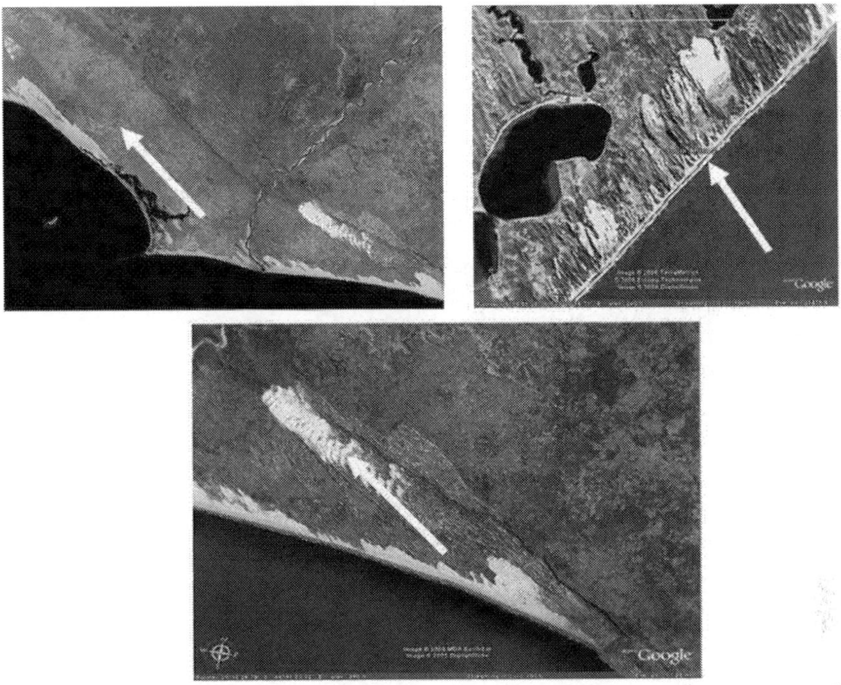

Author's depiction of chevron dunes showing direction of driving force. Note: bottom photo is of the Fenambosy chevron with a maximum run-up of 205 m (673 feet) running inland 35 km. Source map: Google Earth

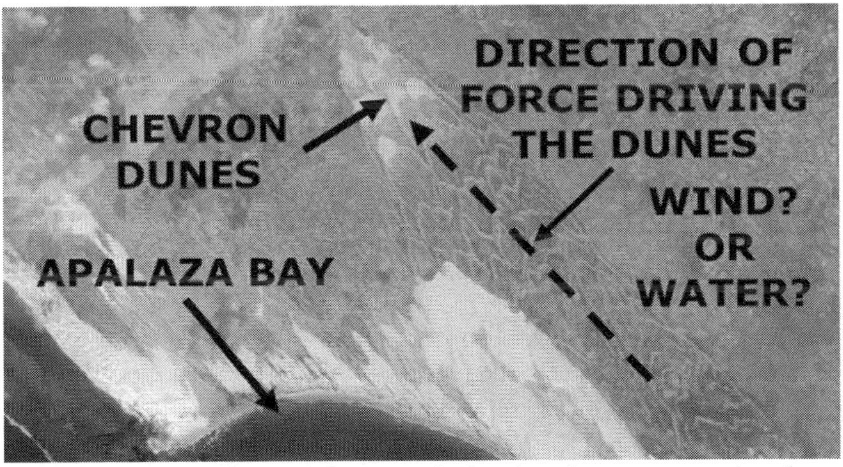

Author's depiction of driving force for dunes near Apalaza Bay. Source map: Google Maps

MASSE'S HYPOTHESIS SUMMARIZED

- Chevrons in southern Madagascar were caused by a mega-tsunami resulting from a major landslide or an impact
- Myths from Greece, Mesopotamia, and Taiwan indicate flood storm came from the south
- A study of comet impact craters by the Holocene Impact Working Group found a candidate crater in the Indian Ocean
- Hypothesis: On May 10, 2807 BC, a three-mile-wide ball of rock/ice created a crater 18 miles in diameter and 12,500 feet below the Indian Ocean surface
- It is known as the Burckle Crater and is located at 30.865°S 61.365°E
- It sent 600-foot-high tsunamis against the world's coastlines
- The tsunami created the Madagascar chevrons and flowed up the Persian Gulf causing Noah's Flood
- It sent plumes of superheated vapor and aerosol particles into the atmosphere
- Hot rain fell out of the sky
- Material ejected into the atmosphere plunged the world into darkness for a week
- Energy from the impact ejected debris to a distance of approximately 9,000 km (5,500 miles)
- According to biblical scholars this event occurred in 2254 BC

While Masse's hypothesis seems to fit the observations of the legends we have reviewed and it predicts a date compatible with the Old Testament, several scientists disagree. Among the leading opponents is Dr. Joanne Bourgeois, a sedimentary geologist at the University of Washington in Seattle, Washington. She and geologist Robert Weiss, of Virginia Tech, assert that it was winds rather than water that formed the chevron dunes. They point out that the chevron dunes line up perfectly with the prevailing winds in that region. Also, they argue, when a tsunami approaches land, it tends to line up with its approach vector normal (perpendicular) to the beach rather than parallel. They do, however, have a problem explaining how rocks and shells can end up in the dune.

Waves (and sand ripples) parallel to the beach with the approach vector normal to the beach. Credit: Dr. James P. McVey, NOAA Sea Grant Program

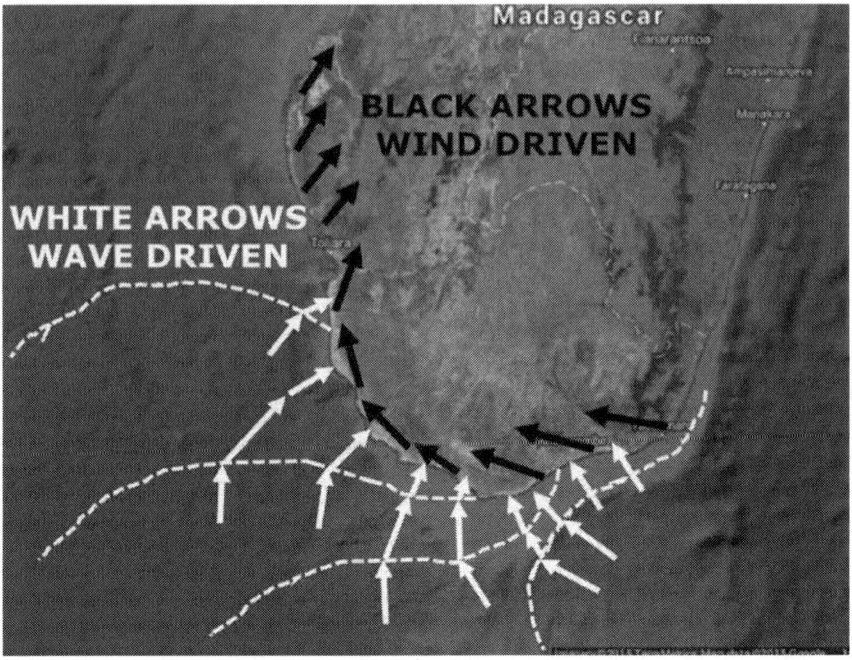

Author's representation of a computer generated model by Robert Weiss comparing wind versus waves as the cause of the dunes. (From similar map by Weiss.)
Source map: Google Maps

Compare the black and white arrows in the figure above with the line between the impact crater and the location of the chevrons in southern Madagascar in an earlier figure.

Before we proceed, we will look closer in the next two chapters at the phenomena known as tsunami and mega-tsunami as well as the rise and fall of sea level associated with the ice age. This background information will factor into other hypotheses about Noah's Flood.

Chapter 9
Tsunamis and mega-tsunamis

Wave energy can be applied to a body of water in many ways. Energy is applied if there is an upshift of the sea floor during an earthquake. Energy is applied if there is an impact by a large object such as a meteor. A large slide of land or ice into water can also impart significant energy. While the wave height at the initiation site may be small, this energy is transmitted through the water at high velocity until it reaches a shore whereupon a wave is generated which climbs up onto the land. This may be a huge wave or a huge surge of water. Below is an illustration.

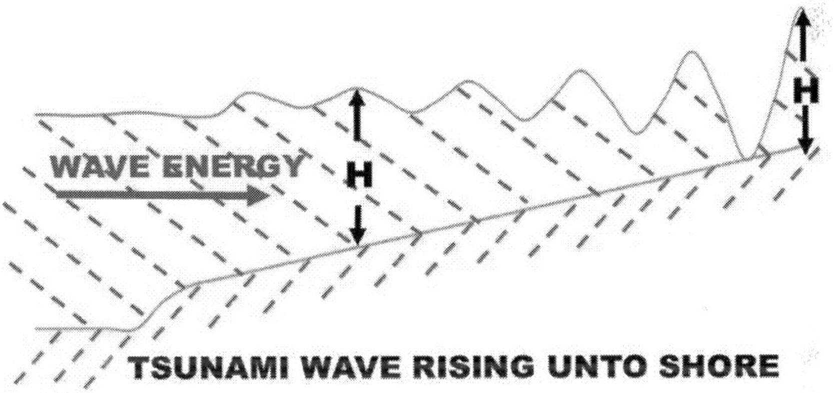

Author's depiction of a tsunami wave. Note wave height relative to the ground remains fairly constant.

The wave travels differently with different depths of the water. In deep water it can travel at 500 to 1,000 km/hr (300 to 600 mph) and creates only a bump in the surface. In shallow water it slows down to only a few mph but can rise to over 30 meters (100 feet).

As the tsunami approaches shallow waters it rises up and may crest if large enough. Credit: Ilhador, Wikimedia Commons

Recent tsunamis which resulted from earthquakes are the Sumatra-Andaman earthquake on December 26, 2004, and the Japan earthquake on March 11, 2011. On December 26, 2004, a 9.0 magnitude earthquake struck off the west coast of Sumatra, Indonesia, resulting in a tsunami that traveled the Indian Ocean at perhaps 500 miles per hour. The effects displaced millions of people and produced about a quarter of a million deaths. This earthquake was caused when the Indian Plate was subducted by the Burma Plate. The entire planet was sent into vibration of about half an inch and quakes were triggered as far away as Alaska. The energy released was 1,500 times that of the Hiroshima atomic bomb.

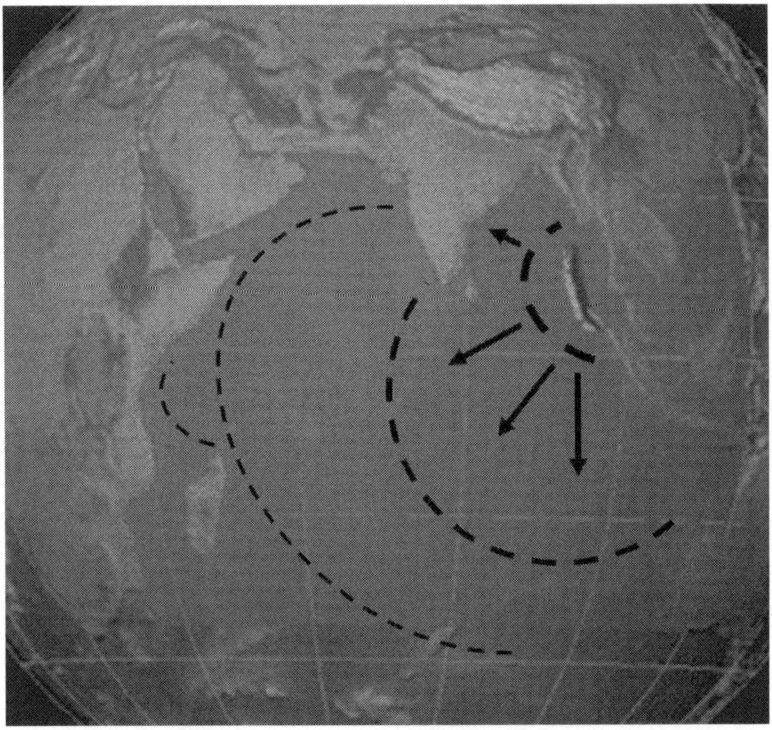

Author's depiction of 2004 Indian Ocean tsunami as it travels outward from Sumatra. Source map: Google Earth

Below is a photo of the wave striking the shores of Thailand.

Wave striking at Ao Nang in Krabi Province, Thailand. Credit: David Rydevik, Stockholm, Sweden, Wikimedia

On March 11, 2011, a major 9.0 earthquake occurred off the coast of Miyagi, Japan. The resulting tsunami reached a maximum run-up of 127 feet near the town of Miyagi. This was the highest ever recorded in Japan. It also caused a 20 to 30 foot surge of water to engulf several towns and kill more than 19,000 people. Over 150,000 homes were damaged or destroyed and serious damage was done to the Fukushima nuclear reactor. Notice in the photo below how the tsunami manifested as a giant surge.

2011 Japan Tsunami. Note: It is more of a big surge than a big wave. Credit: Photo/ Mainichi Shimbun and Tomohiko Kano

Tsunamis are bad enough, but the Earth is also plagued by even larger events called mega-tsunamis which have waves much higher than tsunamis. Wave heights of 500 to 1,000 feet would not be unusual for a mega-tsunami.

On July 9, 1958, in Lituya Bay a mega-tsunami occurred as a result of 40 million cubic yards (a cube over 1,000 feet on a side) of mountainside blowing out and falling 3,000 feet into the bay. The release of energy was so great that near the landfall, trees were destroyed that were over 1,720 feet above sea level. The waves traveling down the bay destroyed trees 600 feet above the lake.

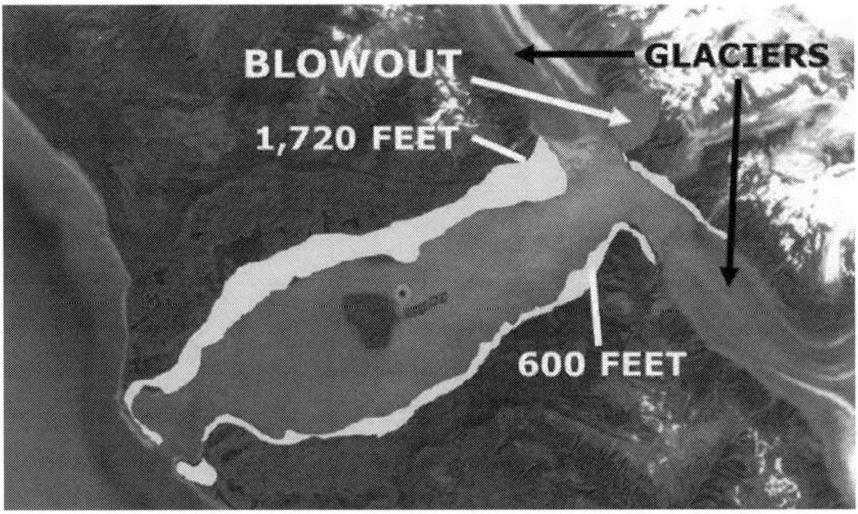

Author's depiction of Lituya Bay mega-tsunami from blowout destroying trees (lightest color) up to 1,720 feet above sea level. Note glaciers flowing into the eastern end. Source map: Google Maps

Two other incidents in recent times are a 1963 event in Italy and a 1980 event at Mt. St. Helens. In 1963 an enormous slab from Monte Toc, a mountain near Venice, Italy, slid into the reservoir at Vajont Dam. Traveling at an estimated 70 miles per hour, it created a surge wave over 850 feet high. In May 1980, the upper 1,500 feet of the former Mt. St. Helens slid into Spirit Lake sending a wave 820 feet high around the lake.

Chapter 10
Melting ice and rising sea level

The Earth has gone through many ice ages over the past few million years. As the ice builds up, sea level drops only to rise again when the ice melts. In the 1930s the Serbian mathematician Milutin Milankovitch recognized a pattern based on the combined effects of three different variations in the Earth's orbit around the sun. These became known as the Milankovitch cycles and are shown below.

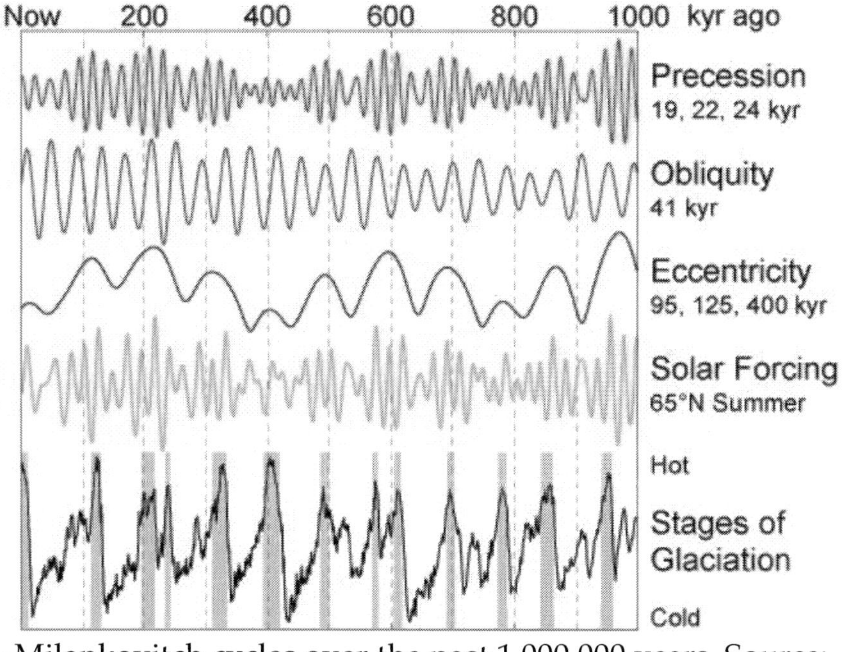

Milankovitch cycles over the past 1,000,000 years. Source: Global Warming Art

The three factors are:

1. The change in the eccentricity of the orbit every 100,000 years caused by the influence of the other planets.
2. The change in the angle of tilt (obliquity) of the Earth's axis every 41,000 years as it varies from 22.1° to 24.5°.

3. The precession of Earth's axis, which changes every 26,000 years.

As the Earth goes through these cycles, the amount of solar radiation (insolation) absorbed by the Earth and the length of the seasons vary. The radiation received during July at a latitude of 65°N is usually used as an indicator and correlates well with the Milankovich cycles. It has been shown that the changes in ice buildup and in sea level also correlate well with these cycles.

There are three major glacial contributors to sea level variation: Fennoscanadian, North American/Canadian, and Antarctic. Below is a graph of sea level over 140,000 years showing how each contributes to sea level changes.

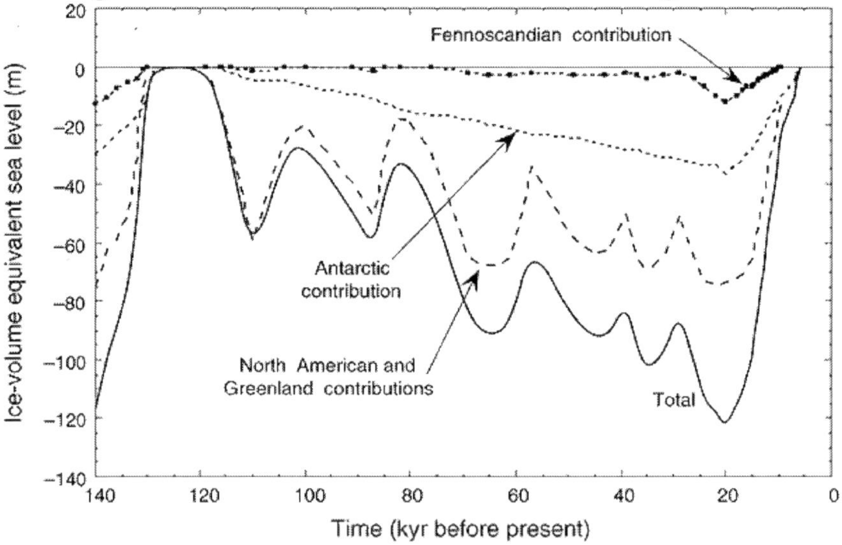

140,000 years of changing sea level. Credit: Climate Change 2001: Working Group I: The Scientific Basis, Intergovernmental Panel on Climate Change. (From Kurt Lambeck, 1999)

Note that the Antarctic ice did not contribute much variation until the end of last ice age 20,000 years ago. Noah's Flood was sudden and not gradual. Assuming Noah's Flood occurred as the last ice age ended, we will focus on sea level changes during the last 20,000 years. Below is a graph of measured sea level changes known as the Post-Glacial Sea Level Rise.

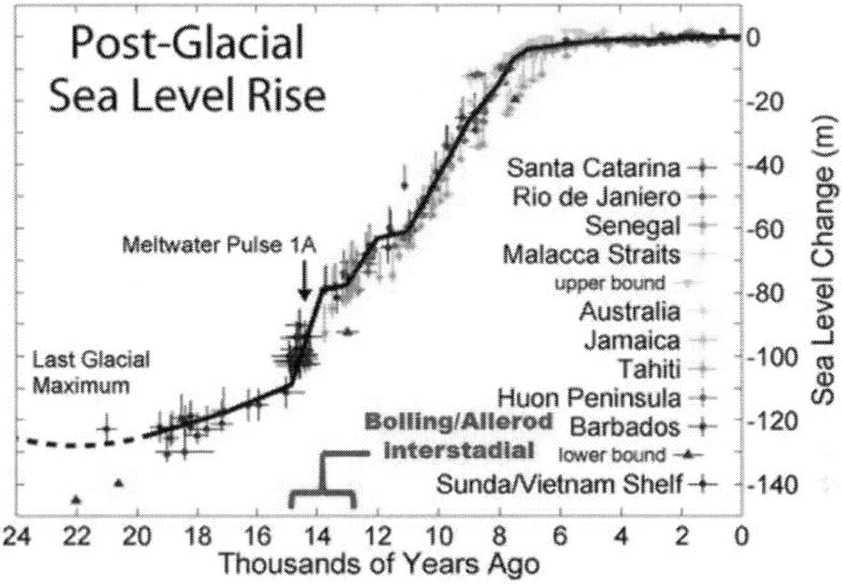

Changes in sea level since last glacial maximum
Source graph: Robert A. Rohde, Global Warming Art project (Wikimedia Commons). Source data: Fleming et al. 1998, Fleming 2000, and Milne et al. 2005

20,000 years ago the Earth was at the peak of the ice age. Much of the northern hemisphere was under ice, and Antarctica had over two times the ice as it does today. The map below shows the northern hemisphere capped with ice.

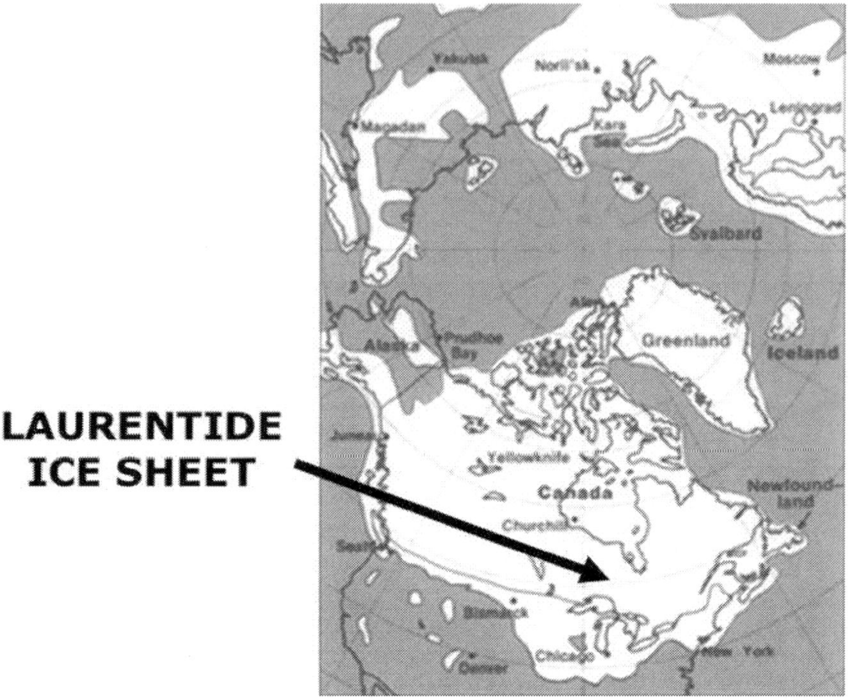

Laurentide Ice Sheet covering northeastern part of North America 20,000 years ago. It was up to 3,000 m (~ 2 miles) thick. Credit: United States Geological Survey

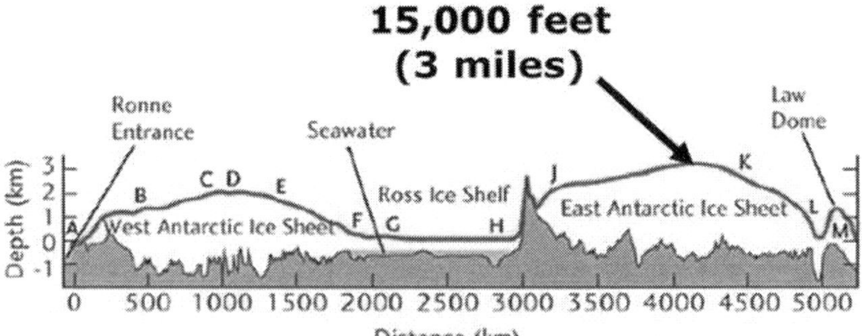

Cross section of Antarctic ice sheet today. Note: East Antarctica has much deeper ice than West Antarctica. Credit: NASA

Today Antarctica has a land area greater than the United States and Mexico and is covered by an ice sheet with an average thickness of 7,000 feet and a maximum thickness of 15,000 feet. 85 percent of Earth's permanent ice resides there.

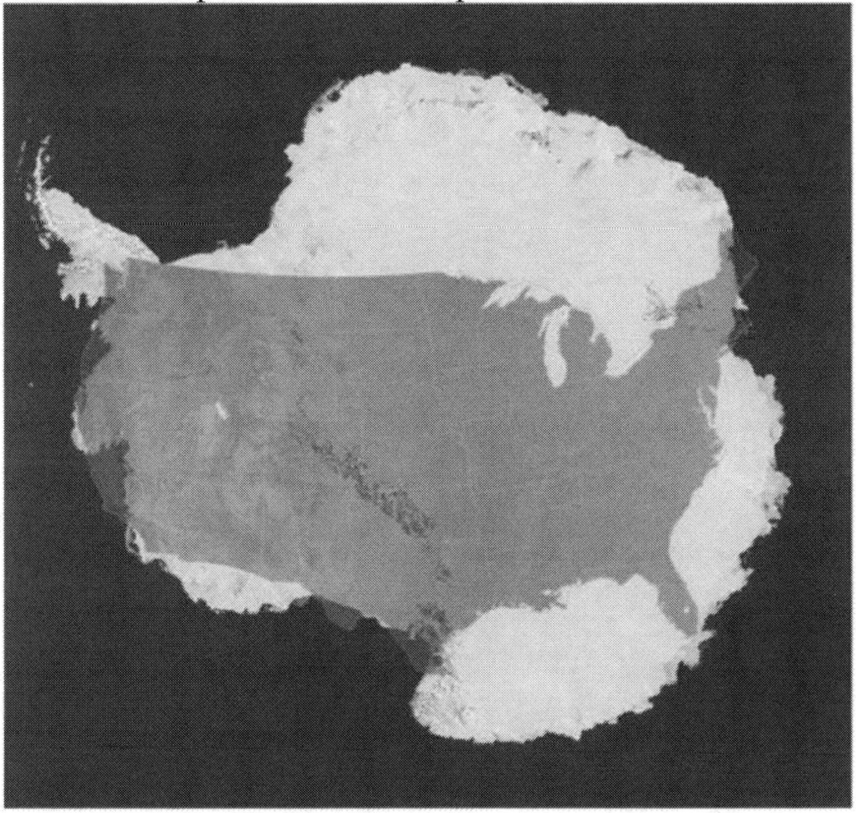

Antarctica compared to the United States today.
Credit: NASA / Google Earth

If all the Antarctic ice were to melt, sea level would rise 120 feet. If all the ice on Earth melted, sea level would rise about 200 feet. The figure below shows what the coastline of the northeastern United States would look like. This map was created by the author using an interactive graphic program at http://geology.com/sea-level-rise/florida.shtml. The sea level was set to 60 meters (200 feet) above present sea level.

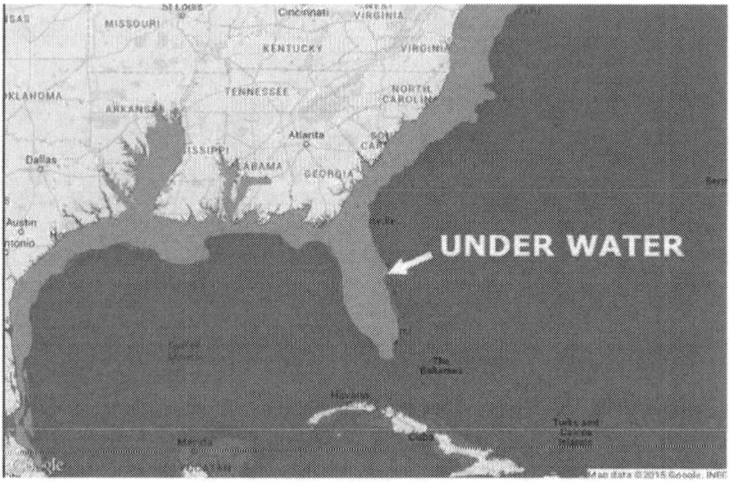

Change in coastline if sea level rises 200 feet. Credit: Author's use of http://geology.com/sea-level-rise/florida.shtml

20,000 years ago the average ice thickness in Antarctica was 21,000 feet, about 4 miles. The maximum ice thickness was 45,000 feet, about 8 miles. That is taller than Mt. Everest! Most of the ice was in the eastern side of the continent. So much water was locked up by ice that world sea levels were nearly 370 feet below present-day. Below is the author's depiction of the ice sheet compared to present.

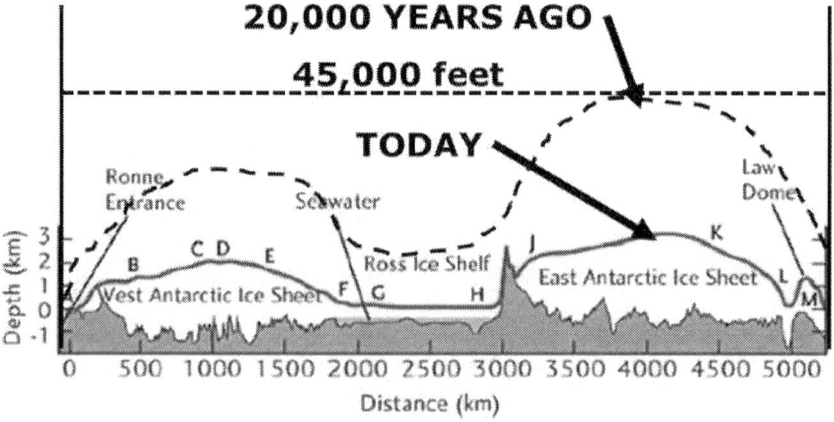

Author's depiction of ice cap 20,000 years ago. Source figure: NASA

Many continental shelves were completely exposed as shown in the NOAA computer simulations shown below.

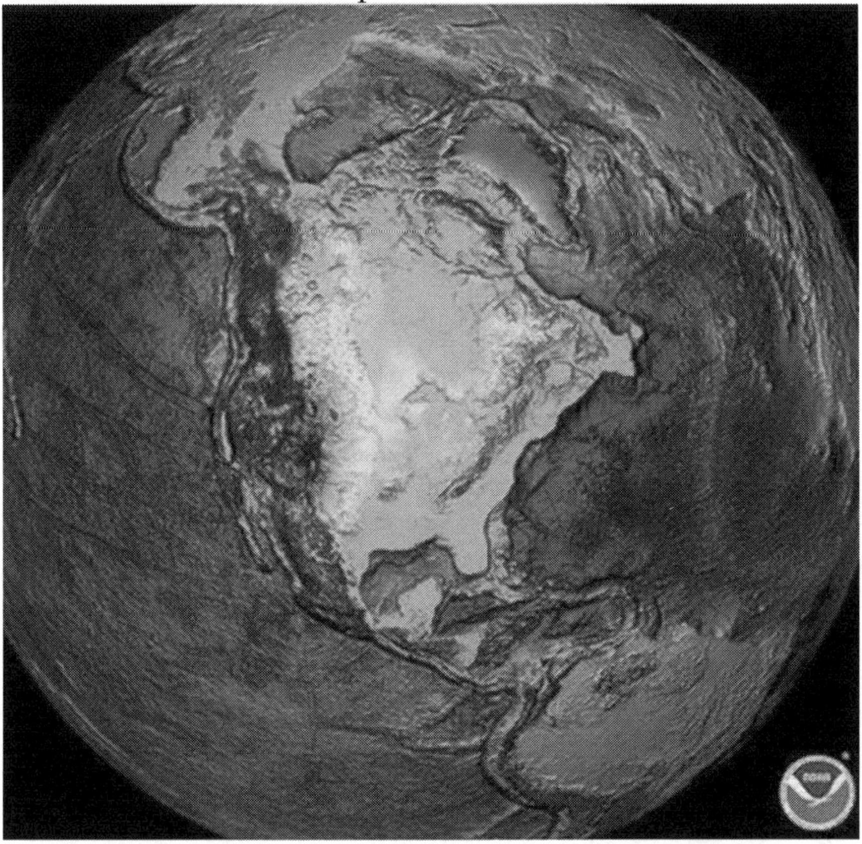

North America 20,000 years ago. Credit: NOAA

Europe and Africa 20,000 years ago. Credit: NOAA

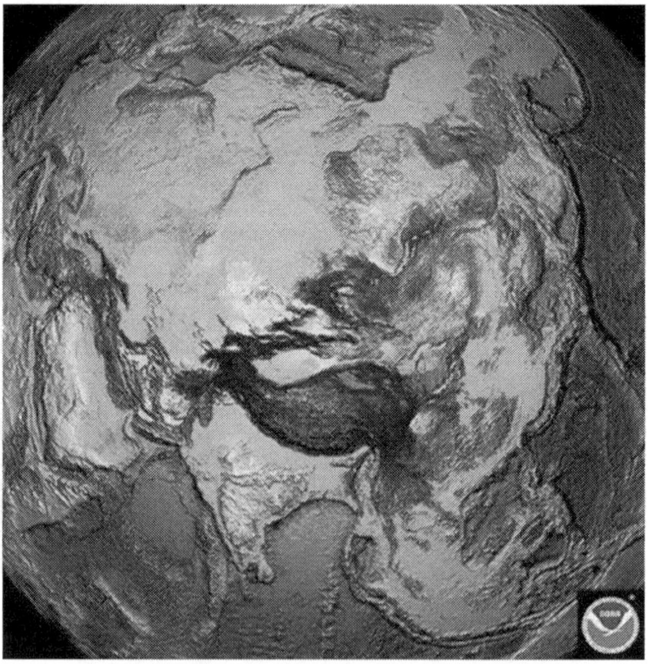

Asia 20,000 years ago. Credit: NOAA

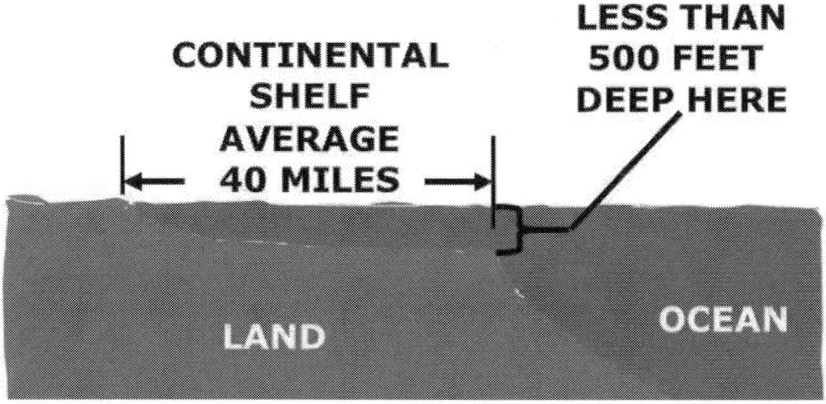

Author's depiction of continental shelf cross section.

The figure above shows how the ocean drops off rapidly beyond the continental shelf. Any settlement located at the edge of the continental shelf 9,500 years ago is now under water. Below is a map showing the location of a sunken city recently discovered in the Gulf of Cambay. It is 120 feet below sea level and has been dated to be over 9,500 years old.

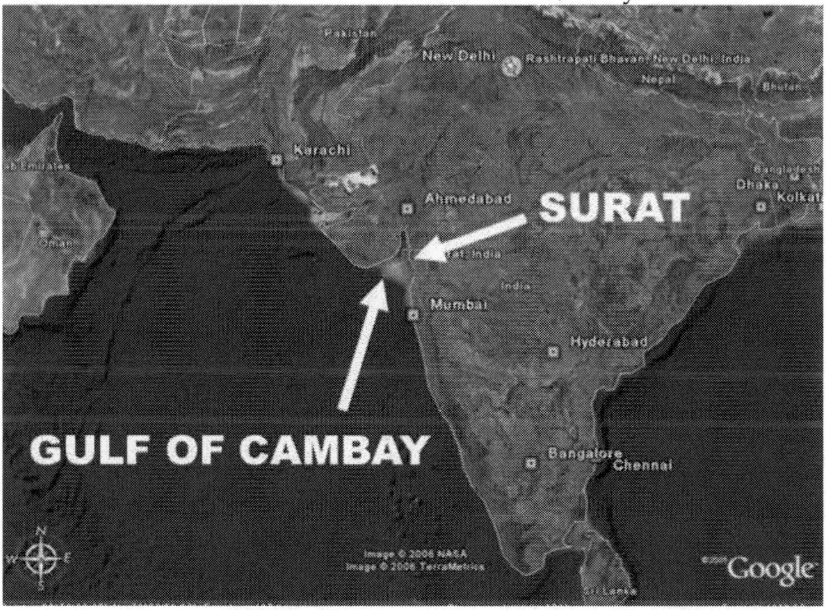

Location of 9,500 year old city under 120 feet of water.
Source map: Google Earth

While building communities on the continental shelf turned out to be a mistake, exposed continental shelves did make human migrations easier. Humanity seems to locate near the mouths of rivers that empty into the sea. Here there is a source of sea food and fresh water. Any human establishment that grew up 15,000 to 20,000 years ago along the exposed ancient sea coast is now under over 200 to 300 feet of water. Much more of our past is probably hidden under water at the edges of continental shelves. There is much work ahead for the new field of underwater archeology.

Any ice supported by land will impact sea level if it melts or slides into the sea. Surrounding Antarctica are numerous ice shelves where the ice extends out from the land and is supported by water. If this ice melts, there is no effect on sea level as the ice is already in the sea. However, ice shelves do retard the flow of glaciers into the sea. As an ice shelf calves icebergs closer and closer to the grounding line, the glacier behind it flows faster and faster toward the sea. As the Earth warms up, this process can occur rapidly.

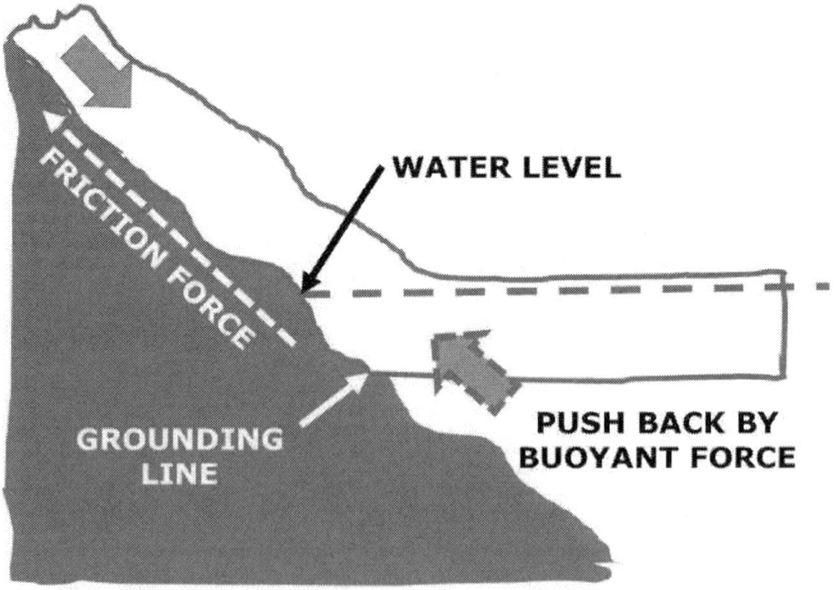

Author's depiction of a stable ice shelf and glacier.

In the figure above, the stable glacier flows slowly toward the sea as its flow rate is held in check by friction and the buoyant push back force from the ice shelf floating on the sea ahead of it. During a warming spell the ice shelf starts to break up and offers less resistance on the glacier. In the figure below we see that eventually the ice shelf breaks up back to the grounding line. The glacier moulins add lubrication to the bottom of the glacier and without the retarding force from an ice shelf, it slides rapidly into the sea.

Surface water flows down into a "moulin" shaft.
Credit: NASA via Wikimedia Commons

Scientists lower a probe into a moulin in Greenland. The probe was lowered to the bottom of the ice sheet.
Credit: NASA

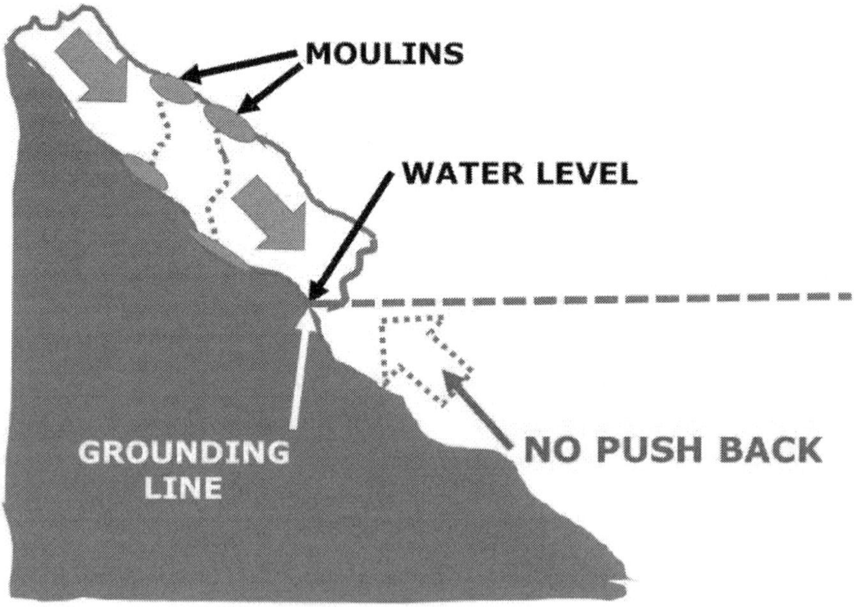

Author's depiction of an unstable ice shelf and glacier.

In 2002 the Larsen B Ice Shelf, located on the western edge of Antarctica, collapsed (broke up) in a matter of about ten weeks, much faster than predicted. The area involved was about the size of Rhode Island. This illustrates how quickly things can change during global warming. The two photos below taken from NASA's Terra satellite document this event.

Larsen B Ice Sheet on January 31, 2002. Credit: NASA

Larsen B Ice Sheet April 13, 2002. Note: Area of collapse is similar to the state of Rhode Island. Credit: NASA

Although ice shelves can collapse rapidly, they have little effect on sea level and by themselves their melting ice would not be the cause of Noah's Flood. Let's look for other possible causes.

Chapter 11
Sitchin's hypothesis refined

Armed with information presented in the previous chapters, it is time to present an hypothesis about the event which led to the story of Noah's Flood. The assumption is that the event occurred in Mesopotamia. The logic behind this is that Abraham was born in Mesopotamia, either in Ur, a city in southern Iraq as Woolley said, or in Urfa, a city in southeastern Turkey. Abraham grew up in the region where the Epic of Gilgamesh would have been a popular story. At some point, Terah, his father, took the family to Harran. Later, Abraham migrated to Canaan and founded the Hebrew religion. It would be natural for Abraham to carry the Epic of Gilgamesh with him as his family migrated from the city of Ur (or Urfa) into Canaan.

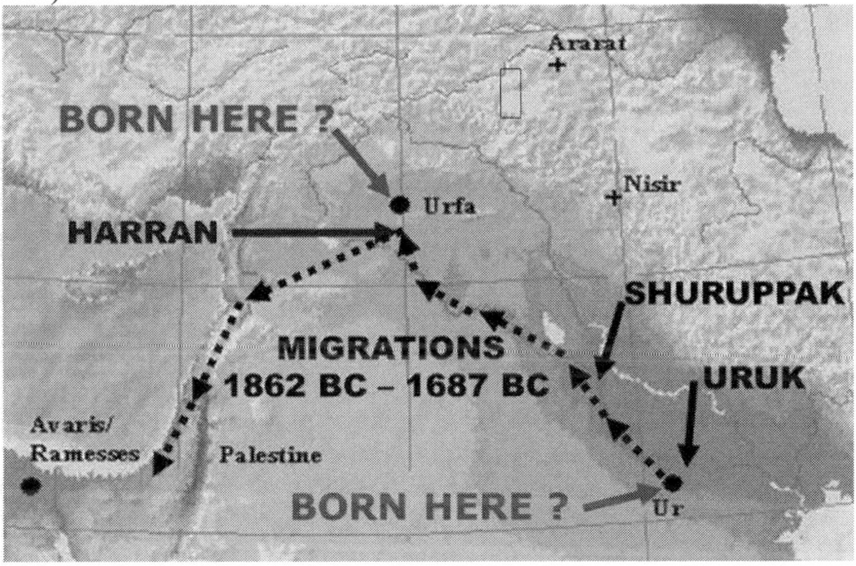

Author's depiction of Abraham's birthplace and migration assuming he was born in southern Messopotamia.
Source map:
http://www.mideastnews.com/kuwait26.html

As mentioned at the beginning of this book, the genesis for this book was Sitchin's flood hypothesis. The Earth began to exit the ice age about 20,000 years ago. He believed that the deluge, as described in the ancient epics which preceded the Old Testament, occurred about 13,000 years ago as a result of melting of the polar ice caps. Sitchin believed that the collapsing ice sheets resulted in a mega-tsunami which flooded southern Mesopotamia. He believed that the ice sheets became unstable as the planet warmed and during a close approach of Nibiru, its gravitational influence triggered the collapse. In the next chapter the reader will learn about the Earth's exit from the ice age and the possibility of mega-tsunamis resulting from collapsing ice sheets.

In ***The 12th Planet,*** Sitchin spends considerable time dating the close approaches of Nibiru every 3,600 years and the influence they had on human events. He frequently mentions rockets when discussing the Anunnaki's mode of travel. I believe that any beings capable of travel within our solar system probably had much more advanced technology than rockets. Perhaps they used field propulsion as exhibited by present-day UFOs. They could probably travel freely between their home planet, Nibiru, and Earth. Thus, other than the appearance of Nibiru as a bright star in the sky, I do not believe a close approach every 3,600 years would have been that significant. The rest of this chapter will bring forward scientific discoveries over the last 40 years and offer a revised hypothesis to that of Sitchin's work.

REVISED HYPOTHESIS

"After the kingship descended from heaven, the kingship was in Eridug [Eridu]. *In Eridug, Alulim became king; he ruled for 28,800 years."* This quote is from the following website: http://en.wikipedia.org/wiki/Sumerian_King_List

According to Sitchin, there was a relatively small (perhaps 50 individuals) landing party of Anunnaki from the planet Nibiru who came to Earth for the purpose of mining minerals, primarily gold. It was the "lesser gods" of the Anunnaki who performed the mining labor. Sitchin states they took 3,600 years to discover that mining was hard work and proceeded to rebel. The science officer, Enki (think of him as Spock), solved the problem. As it turned out, Homo erectus inhabited the planet at that time. In order to create a more intelligent worker-being capable of communication, Enki used genetic engineering to combine the Homo erectus genes with their own. Homo sapiens was born. Unlike our mules, they were eventually "designed" to be able to reproduce. Also, they were "designed" to have minimal aging. After all, why create a worker who wears out? In the future, these two features would create an overpopulation problem. Spock would have to deal with this mistake in a few hundred millennia.

According to Sitchin, Homo sapiens were created about 300,000 years ago. In Allan Wilson's study "mitochondrial Eve" was not the only woman alive at the time. She was just the most recent woman to possess the mtDNA from which all existing Homo sapiens' mtDNA derived. Thus, according to Allan Wilson's study, the first human was created about 200,000 years ago.

When did the Anunnaki arrive on Earth? Sitchin said they landed about 445,000 years ago. According to the tablet giving the kingship before the Deluge, the first seven "gods" ruled for 221,800 years after landing. Kingship was moved to Shuruppak where Ubara-Tutu ruled for 18,600 years before the flood. That would place the flood 240,400 years after the Anunnaki landed. I will present an hypothesis that the flood occurred 14,200 years ago. That would place their landing at about 255,000 years ago.

Sitchin's estimate of 300,000 years for the creation of humans is probably too high. If the Anunnaki created humans after "lesser gods" worked 3.600 years before rebelling, then humanity began about 441,000 years ago. In a 2009 article in the *American Journal of Human Genetics*, Pedro Soares provides a range of 99,000 to 234,000 years BP as an estimate for the creation of humanity. It would make more sense to place the creation of Homo sapiens at about 250,000 years ago. This would place the arrival of the Anunnaki at about 255,000 years ago instead of 445,000 as Sitchin believed. The figure below is based on 250,000 BC for when Homo sapiens appeared in Africa.

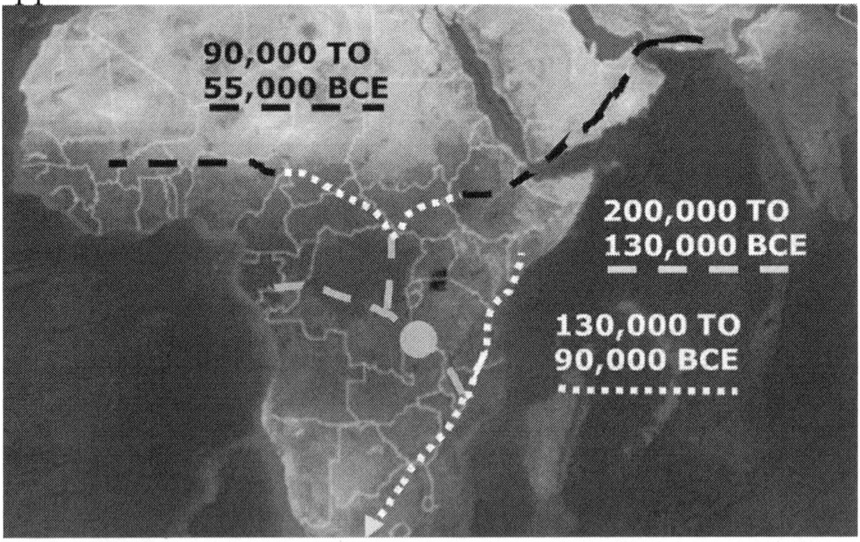

Author's depiction of human migration out of Africa. Data based on several studies, Behar et al. 2008, Gonder et al. 2007, Reed and Tishkoff. Credit: Wikipedia

THE FLOOD

According to my hypothesis, 14,200 years ago the Persian Gulf was a lake located 370 feet above sea level. It was only 50 to 150 feet deep and was fed by four rivers. According to Genesis 2:10–14 *"And a river went out of Eden to water the garden; and from thence it was parted, and became into four heads. The name of the first is Pison* [probably the Wadi al Batin or Kuwait River today]*: that is it which compasseth the whole land of Havilah* [Saudi Arabia]*, where there is gold; and the gold of that land is good: there is the bdellium* [a resin similar to myrrh] *and the onyx stone. And the name of the second river is Gihon* [probably the Karun River out of Iran today]*: the same is it that compasseth the whole land of Ethiopia* [probably Iran]. *And the name of the third river is Hiddekel* [Tigris]*: that is it which goeth toward the east of Assyria. And the fourth river is Euphrates."* The Persian lake emptied into the Arabian Sea 370 feet below through rapids in what is now the Strait of Hormuz. It was also headed by extensive marshlands to the north. See the map below.

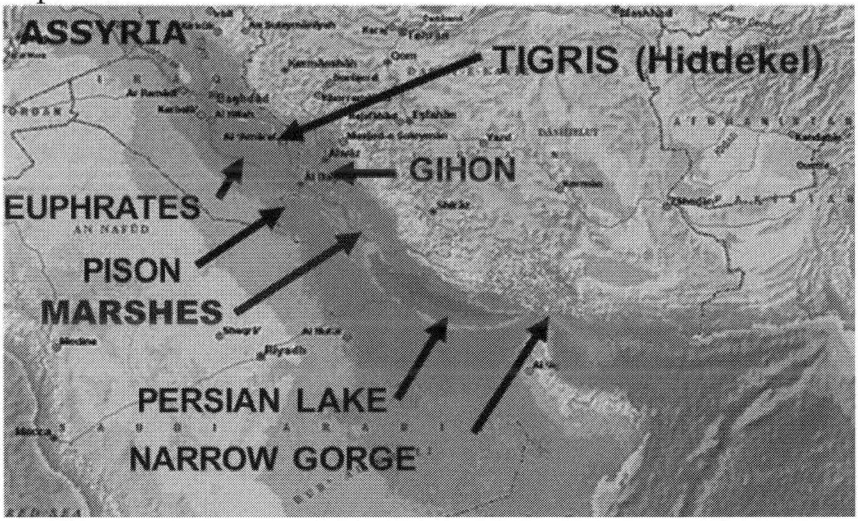

Author's depiction of the Persian "lake" fed by four rivers. Source map: **http://www.mideastnews.com/kuwait26.html**

Above is the author's depiction of the "Persian lake" 14,200 years ago when the sea level was 370 feet lower. The map above shows the possible location of the four rivers mentioned in the Bible. The lake would have emptied down through a gorge into the Arabian Sea 370 feet below.

The planet was warming as it exited the ice age. The ice shelves in Antarctica began rapidly collapsing allowing the massive eight-mile-high glaciers along the east coast of Antarctica to slide down the slopes into the Indian Ocean.

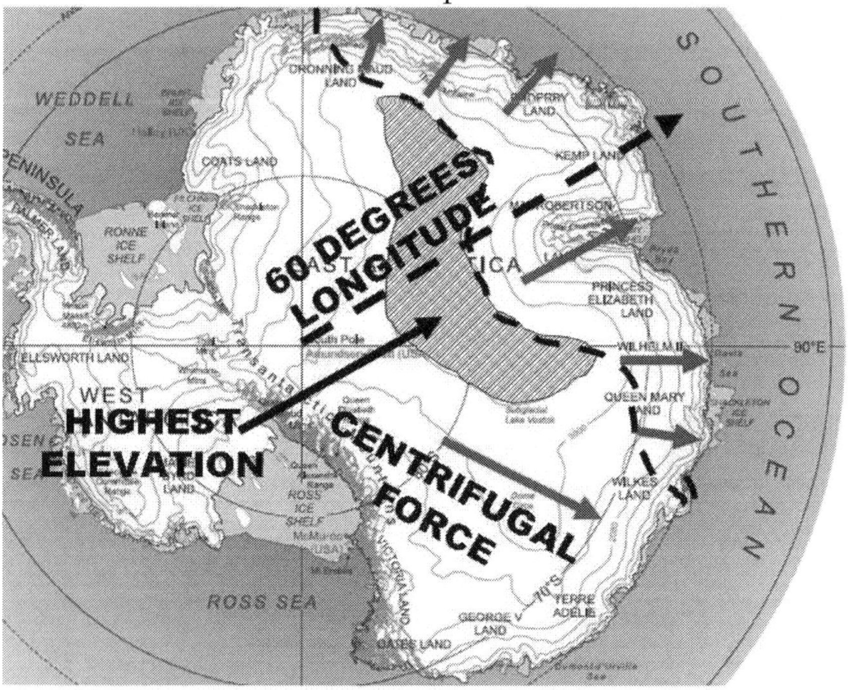

Author's depiction of Antarctica showing topography and forces acting on glaciers facing east along the Southern Ocean, facing the Indian Ocean. Source map: NASA

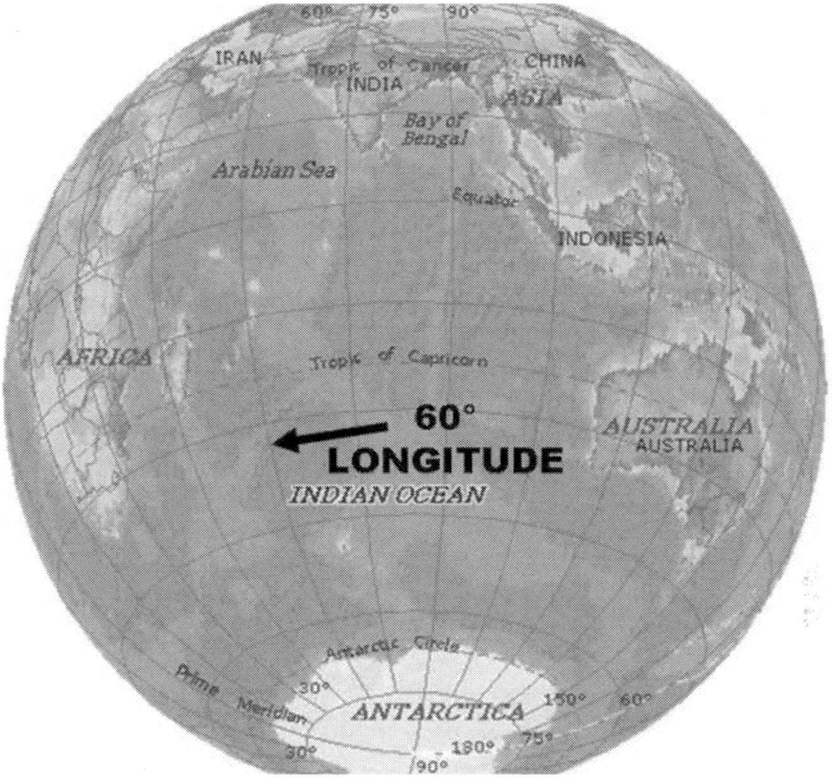

Antarctica relative to Indian Ocean. Credit: NASA

Hundreds of cubic miles of ice were suddenly released and fell thousands of feet as the glaciers slid down the eastern slopes of Antarctica. In the figure above note that most of the ice mass is east of the axis of rotation. Also, there is a steep topographical gradient along the coast facing the Indian Ocean.

The vibrations set off chain reactions with even more glaciers sliding into the water. The release of such huge amounts of energy shook the earth, triggering earthquakes around the globe. The Zagros Mountains east of the Tigris River also shook and thundered with earthquakes. Noah awaited this signal to board the craft he had constructed in the ancient city of Shuruppak.

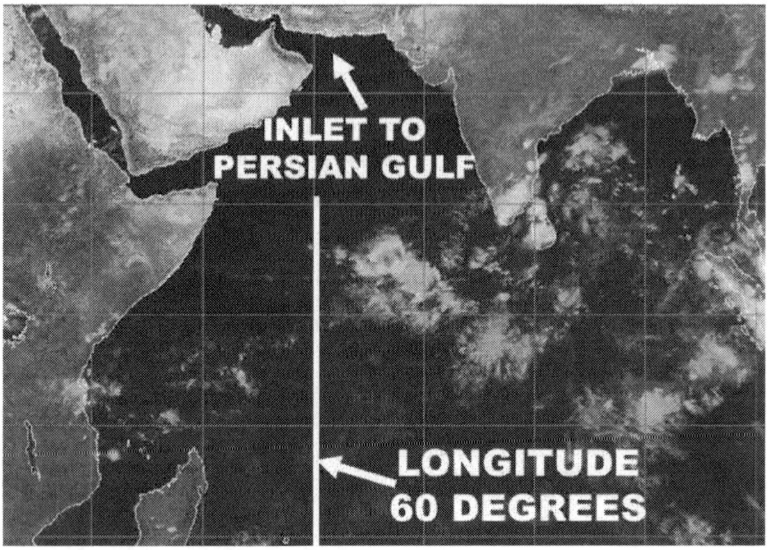

Mega-tsunami heads into Persian Gulf.
Source map: Google Maps

A mega-tsunami traveling up from the south sent huge waves as well as a seemingly unending surge several hundred feet high into the Persian "lake" carrying Noah northward to higher ground where he landed south of the headwaters of the Tigris River, perhaps within sight of Mt. Judi and Mt. Ararat.

Deluge carries Noah north toward Mt. Ararat.
Source map:
http://www.mideastnews.com/kuwait26.html

After such an ordeal Noah would have headed north toward higher ground and settled close to the headwaters of the Euphrates near present-day Turkey. He would have begun civilization again, building a temple to his Lords (gods). His sons would have spread his "seed" in all directions from that location. I will test this hypothesis by presenting more detailed information below.

The question to ask might be, why did I set the date at 14,200 years BPE? Some Biblical chronologists have placed the event at 4,285 years ago. The graph below indicates that 14,200 years ago there was a sudden rise in sea level known as the Meltwater Pulse 1A. The slope of the curve at that time is not well defined with a discontinuity going straight up for 80–100 feet or so. Normally, natural events follow a well behaved curve, so it would be reasonable to assume that sea level rise would do likewise as the planet warmed and melted glacier ice during the Bolling/Allerod interstadial.

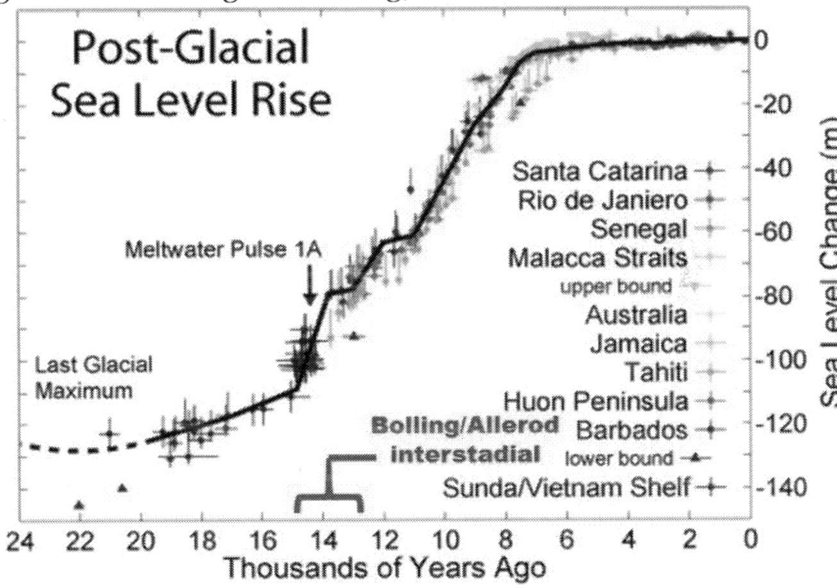

Changes in sea level since last glacial maximum.
Source graph: Robert A. Rohde, Global Warming Art project (Wikimedia Commons). Source data: Fleming et al. 1998, Fleming 2000, and Milne et al. 2005

Above is the graph showing global sea rise since the last glacial maximum. Below are two altered versions of that graph. They show this author's attempt at fitting two smooth curves with a vertical offset to represent the Meltwater Pulse 1A. The required offset is about 25–30 meters (about 80–100 feet).

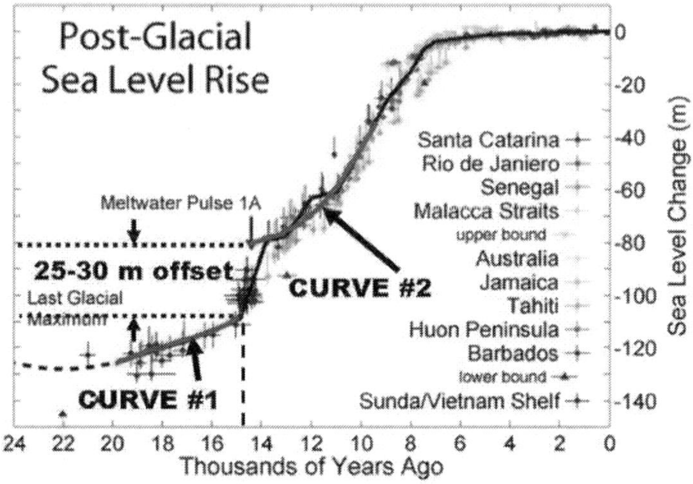

Author's depiction of two curves (#1 and #2) showing well behaved melting. Source graph: (same as above)

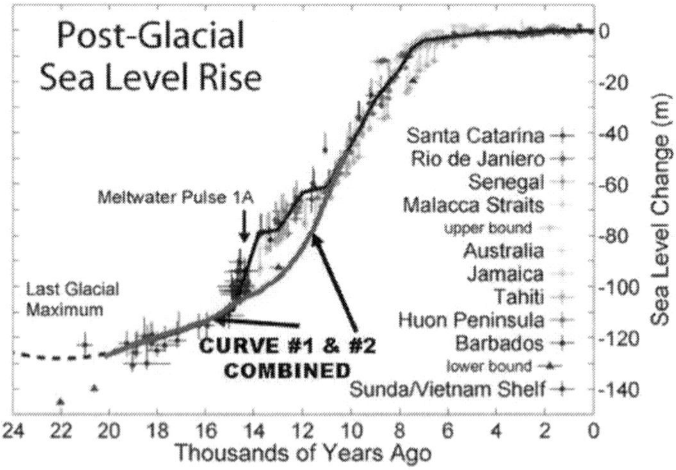

Author's blending of curve #1 and #2 to produce smooth curve. Source graph: (same as above)

Curves #1 and #2 combined, represents about 12,000 years of sea rise history from 20,000 to 8,000 years ago. Below is a plot of that data compared to a third-order polynomial curve. The result shows that the sea rise did indeed follow a smooth curve except for the 80–100 foot discontinuity at Meltwater Pulse 1A. For convenience I will repeat: Dr. Wilson's "theory of ice ages implies that the present interglacial will end with, or at least be interrupted by, an Antarctic ice sheet 'surge.' Such surges in the past would have caused distinctive rises of sea level by 10–30 m, in 100 years **or much less** [emphasis added by author], and precisely at the break of climate at the end of each interglacial." The graph above seems to support that theory. Taking the data from the composite graph and replotting it results in the graph below.

SEA RISE BEGINNING 20,000 YEARS AGO FIT BY 3RD ORDER POLYNOMIAL h = 0.0889 yr^3 -0.7889 yr^2 +4.3 yr - 1.6

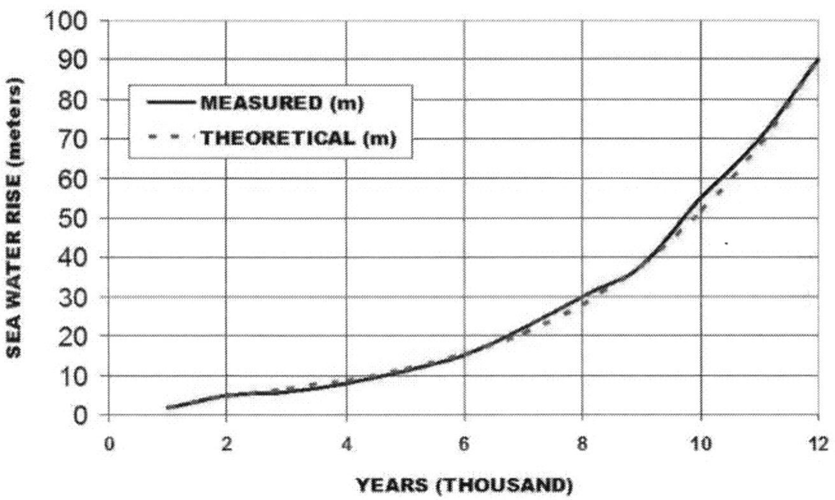

Author's fit of sea rise data to a polynomial curve.

The increase in slope of the curve as sea level rises can be explained by three factors:

1. Increase in solar insolation as seen in the earlier graph of a plot of Milankovitch cycles.

2. Change in albedo as melting ice exposes more land.
3. Increasing surface area of the world's oceans as the continental shelves are covered by water. See below.

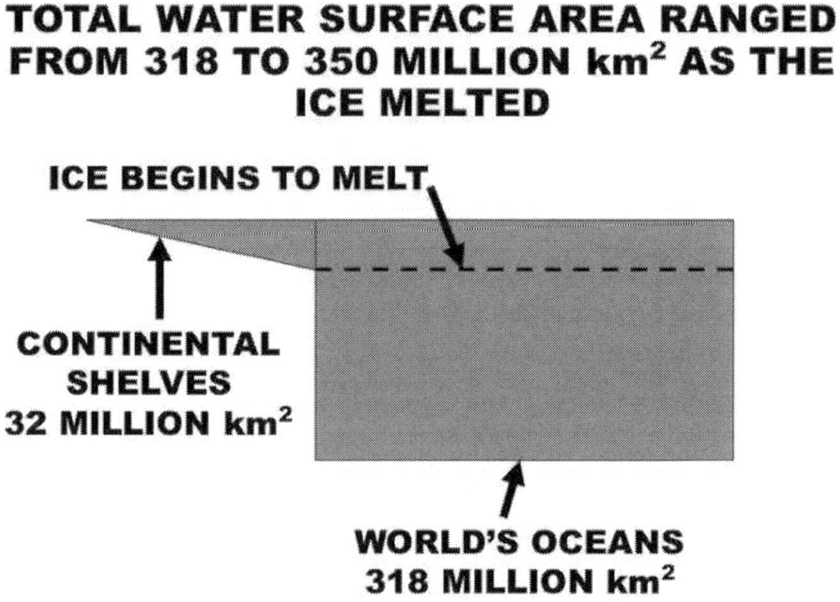

Author's illustration of increase in ocean surface area as continental shelves are flooded as sea is rising.

From the above illustration, it can be seen that surface area of the world's oceans increases about 10 percent as continental shelves flood during ice melt. It seems reasonable to fit Noah's event to a sudden rise of about 80–100 feet in sea level 14,200 years ago. See Appendix A for a discussion on fitting curves to sea rise data and contributing factors such as solar insolation.

Earlier, the claim was made that the Persian Gulf was quite shallow and would have been a freshwater lake 14,200 years ago. The map below shows the present depth of the Persian Gulf. Note the sharp increase in depth near the mouth of the Strait of Hormuz. This may be due to erosion from a high flow rate of water in the past.

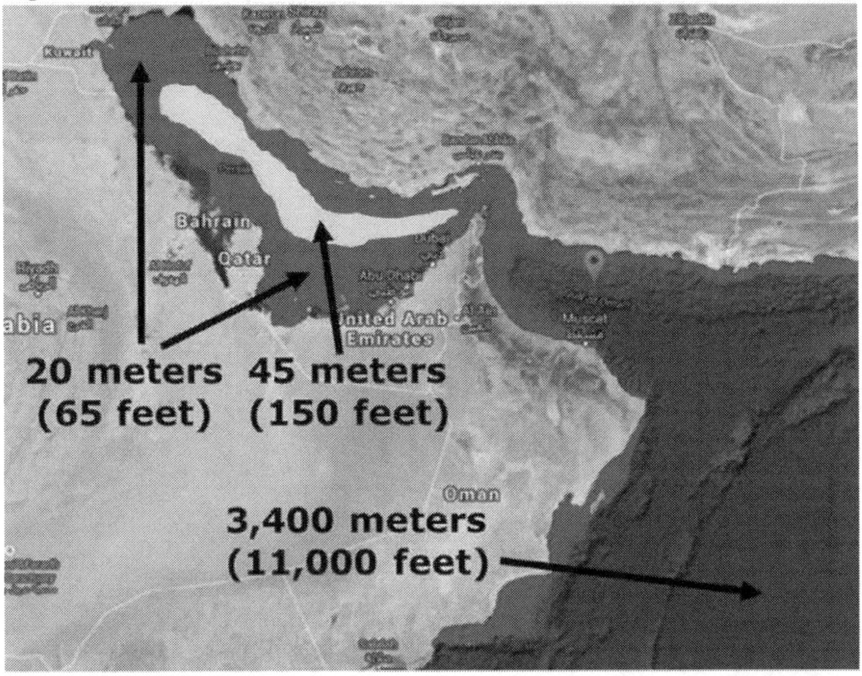

Author's depiction of how shallow the Persian Gulf is today. Source map: Google Maps

John T. Hollin, Ph.D., is a Fellow Emeritus of the Institute of Arctic and Alpine Research (INSTAAR) at the University of Colorado. In 1972 he made reference to glaciologist Professor Allan T. Wilson's work at Victoria University in New Zealand. Hollin stated, "Wilson's theory of ice ages implies that the present interglacial will end with, or at least be interrupted by, an Antarctic ice sheet 'surge.' Such surges in the past would have caused distinctive rises of sea level: by 10–30 m [30–100 ft.], in 100 years or **much less** [bold added by author], and precisely at the break of climate at the end of each interglacial."

Looking at the topographical map below, it can be seen that a combination of distance from the axis of rotation (centrifugal force) and a steep topographical gradient along the coast facing the Indian Ocean would make that area most likely to slide into the ocean. Note also that most of the ice mass is east of the axis of rotation.

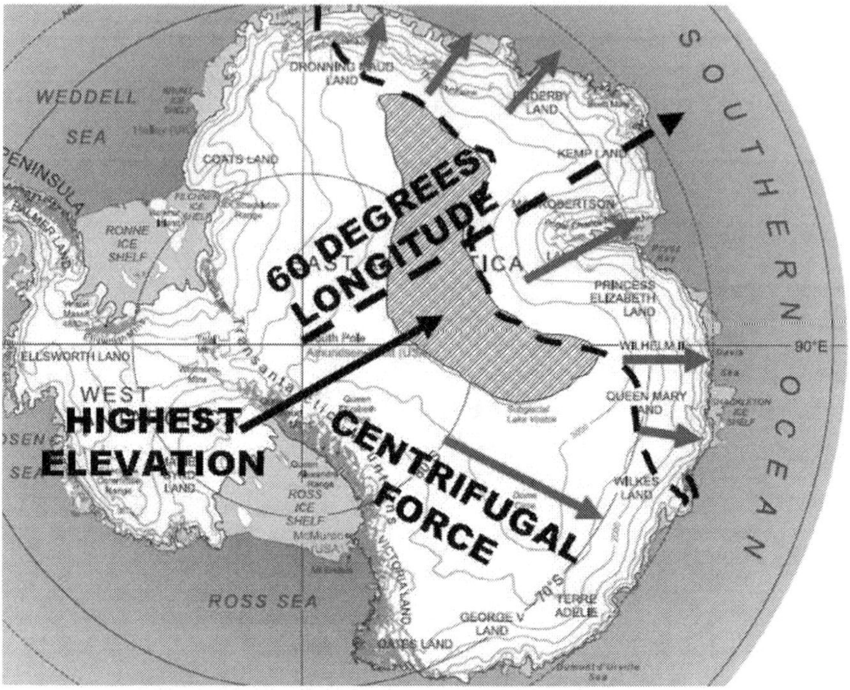

Author's depiction of Antarctica 14,200 years ago as eight-mile-high glaciers are ready to slide into the Indian Ocean. Source map: NASA

What triggered this major slide 14,200 years ago? Did the Earth's crust shift as Hapgood suggested? Was there a major meteor impact in Antarctica? Sitchin has speculated that the rogue planet, Nibiru, made a close approach to Earth, and its gravitational influence was the trigger. This author discounts that hypothesis (Appendix B). Simple calculations show that the Moon's gravity at its closest approach is 26 percent higher than at its farthest. The size of Nibiru has not been established, but even if it is as large as Jupiter, when it is in the asteroid belt and at its closest approach to Earth, its gravitational influence is only 6 percent as much as the Moon.

Most people do not even detect the 26 percent variation from the Moon.

Most likely, major slide was simply the result of rising temperatures as we exited the glacial maximum. The warm period known as the Bolling/ Allerod interstadial lasted from 14,700 to 12,700 years ago. The cause of the warming period is not known but it has been posited that a significant solar outburst at this time would bring about a significant and sharp rise in Earth's temperature.

The surface melting of the ice created moulins which added lubrication to the bottom of the ice sheets. That, combined with the melting and collapsing of the ice sheets which normally hold back the glaciers, led to the rapid glacial slides into the sea.

The energy released into the sea would send a mega-tsunami along the 60-degree longitude, and it would be directed into the entrance to the present Persian Gulf. The energy of the mega-tsunami would be focused into the shallow region known as the Strait of Hormuz and rise up perhaps 500–1,000 feet. A wave perhaps 500 feet high would surge northward following the Tigris-Euphrates valley into southern Turkey.

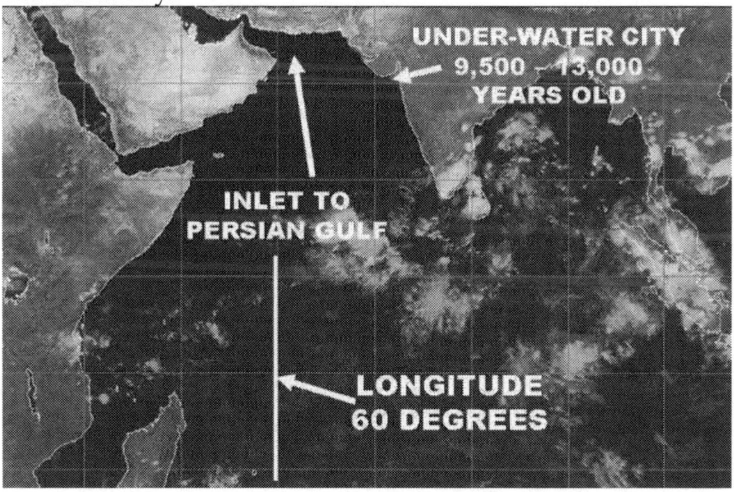

Mega-tsunami directed into the Persian Gulf.
Source map: Google Earth

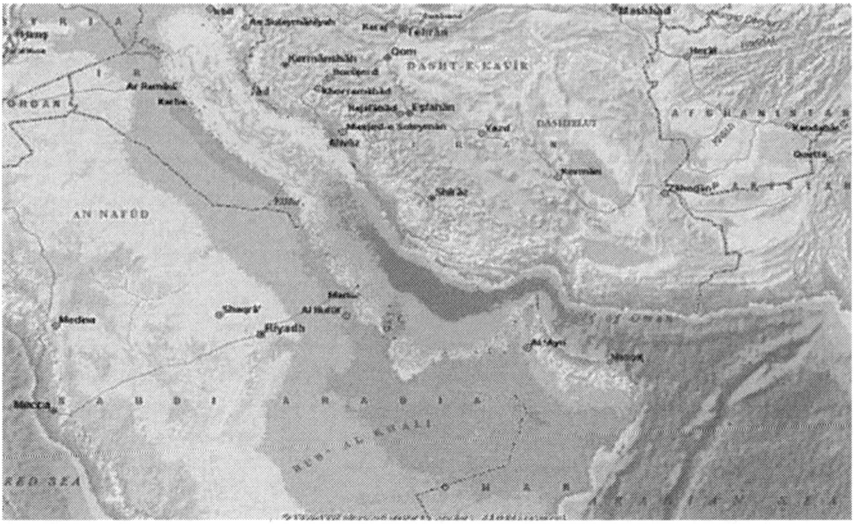

Author's depiction of the deluge in process. Source map: http://www.mideastnews.com/kuwait26.html

The mega-tsunami would have breached the Strait of Hormuz and carried everything northward with its 500 to 1,000 foot high surge of water. As he was traveling with the surge, Noah probably would not be able to see any land as he would have been 50 to 100 miles from the water's edge. Even the 6,500 foot Zagros Mountains to the east and the 17,000 foot peak of Mt. Ararat to the north would be over the horizon. To him, the whole world was under water. As he drifted north and neared his landing site, the peaks of Mt. Ararat would become visible. Without any other point of reference, he would believe the water was receding.

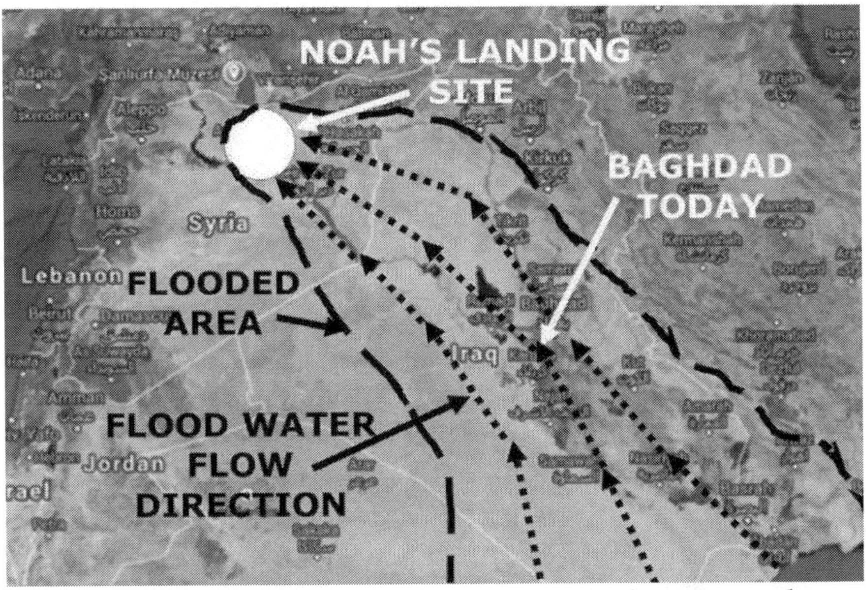

Author's depiction of water surge which carries ark northwest to higher ground. Source Map: Google Maps

The map above shows the water surge from the tsunami carrying Noah's Ark from an elevation of about 120 feet above present sea level to an elevation of 700 to 1,000 feet above present sea level as it travels northwest 400 miles from Shuruppak (Baghdad). The map below is of a larger region showing Noah's probable landing site relative to other points of interest. Harran is shown because of its importance to the history of Abraham. His father, Terah, moved the family there on the way to Canaan. Note the proximity to the area where Gilgamesh found Utnapishtim.

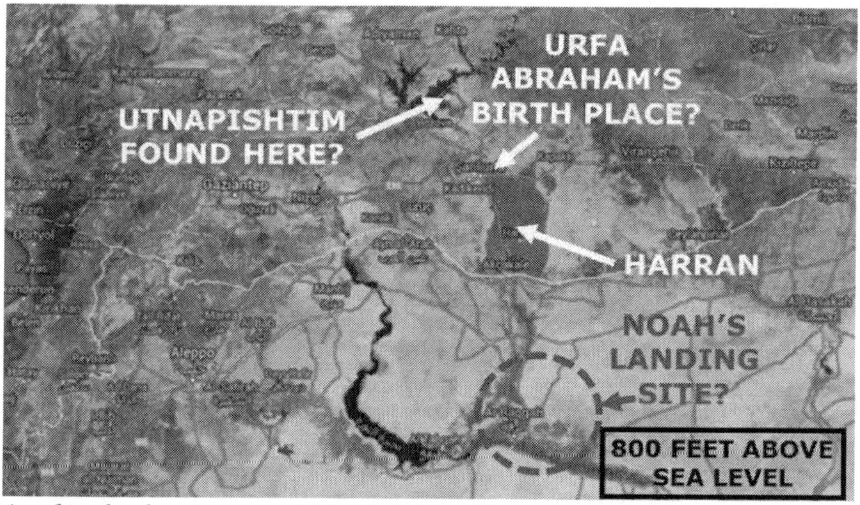

Author's depiction of Noah's landing site relative to Harran.
Source map: Google Maps

Chapter 12
Brief review of archeology

In order to understand what happened before and after the Deluge, the reader must have some understanding of archeology. If the Deluge occurred 14,200 years ago, then Noah was probably a member of the old to middle stone-age people. The following is a comparison of three periods.

PALEOLITHIC (OLD STONE AGE) - 2 million BC to 10,000 BC

1. People were nomadic hunter-gatherers
2. Simple tools and weapons out of stone, bone, and wood
3. Wore animal skins and had a spoken language
4. Lived in caves, built fires, and created cave art (oldest is 40,000 years old)

MESOLITHIC (MIDDLE STONE AGE) - 10,000 to 7,000 BC

1. People learned how to farm
2. They tended to stay in one place
3. They built permanent villages

NEOLITHIC (NEW STONE AGE) – starting at 7,000 BC

1. Domesticated plants and animals
2. Advanced tools and weapons such as calendars, plow, arrow head, and ax head
3. Pottery making and weaving
4. Trade and accumulation of wealth
5. Growth in population

PRE-POTTERY NEOLITHIC A (8000 to 7000 BC)

1. Pottery was not in use
2. Some cultivation of crops with granaries inside dwellings
3. Hunting of wild game
4. Bodies were buried below the floor

5. Dwellings were circular, elliptical, and occasionally even octagonal

PRE-POTTERY NEOLITHIC B (7000 to 6000 BC)

1. Pottery was not in use
2. More dependence on domesticated animals
3. Dwellings were more rectilinear
4. Floors were made of a white lime and clay plaster and were highly polished

What did Noah's Ark look like?

The reader may recall the famous monolog that comedian Bill Cosby did in the early 60s about a conversation between Noah and the Lord. The Lord instructed Noah to build an ark. Noah asked, "What's an ark?" Then the Lord gave Noah the required dimensions in cubits. Noah asked, "What's a cubit?" As funny as the monolog was, it raises an interesting question. If Noah was a late stone-age person, did he know what a boat was and how to build one? The answer is probably yes. Dr. Dennis Stanford of the Smithsonian Institution in Washington, DC, believes 18,000 years ago people had the ability to build boats (umiaks) made of animal skins attached to a wooden frame. They even sailed the Atlantic to the eastern part of North America. Perhaps Greenland was one of their landing sites. Below is a photo of a similar umiak built by the Eskimos in Alaska. Would such a boat survive a tsunami? Perhaps a large kayak with closed hatches on top would survive better.

Eskimos in an umiak on a whale hunt. This is the same ancient design used by the Inuit on the east coast.

Credit: Ted Stevens (R – Alaska) (Public domain)

The world's largest one piece kayak. Credit: http://www.kayak.spirithawk.net/images/large-kayak.jpg

In 1948, a cuneiform tablet was purchased by Leonard Simmons during his wartime service in Egypt. Later, his son Douglas brought it to the British Museum to have it evaluated. Dr. Irving Finkel of the British Museum was amazed at what was written on the tablet and referred to it as the "Ark tablet" because it describes the construction of the ark as given in the flood story epics of Ziusudra and Atra-Hasis. Dr. Finkel's translation of the tablet is summarized below:

- It details the Mesopotamian story and construction of the Noah's Ark
- The ark was to be a kuphar (round boat) 220 feet in diameter
- It was to be constructed with 30 ribs (of wood?)
- The frame was to be bound by woven marsh reeds
- The outside was to be sealed with bitumen which was easily available and common practice

Note: The ark was to be a round boat similar to the one described below by Herodotus at a later time when skins were being used to cover the boat instead of woven reeds. This type of boat (coracle) was traditionally used on the Tigris and Euphrates rivers.

The famous Greek historian, Herodotus, described the typical boat used in Babylon in 450 BC when he visited there. When the "Ark tablet" was written, the typical boat then was probably similar to what Herodotus described as given below.

> *"But that which surprises me most in the land, after the city itself, I will now proceed to mention. The boats which come down the river to Babylon are circular, and made of skins. The frames, which are of willow, are cut in the country of the Armenians above Assyria, and on these, which serve for hulls, a covering of skins is stretched outside, and thus the boats are made, without either stem or stern, quite round like a shield. They are then entirely filled with straw, and their cargo is put on board, after which they are suffered to float down the stream. Their chief freight is wine, stored in casks made of the wood of the palm-tree. They are managed by two men who stand upright in them, each plying an oar, one pulling and the other pushing. The boats are of various sizes, some larger,*

some smaller; the biggest reach as high as five thousand talents' burthen. Each vessel has a live ass on board; those of larger size have more than one. When they reach Babylon, the cargo is landed and offered for sale; after which the men break up their boats, sell the straw and the frames, and loading their asses with the skins, set off on their way back to Armenia. The current is too strong to allow a boat to return upstream, for which reason they make their boats of skins rather than wood. On their return to Armenia they build fresh boats for the next voyage."

Note: A *talent* was a unit of measure equal to approximately 57 pounds. Thus, a large boat could have a displacement of about 285,000 pounds and be about 40 feet in diameter. Below is a photo of a kuphar in Baghdad.

A kuphar in Baghdad - 1914
Credit: Andrew Davidson – Wikipedia – Public Domain

Stone-age art

The sophistication of stone-age people can be seen in their cave art shown below. The oldest cave art goes back to over 40,000 years ago.

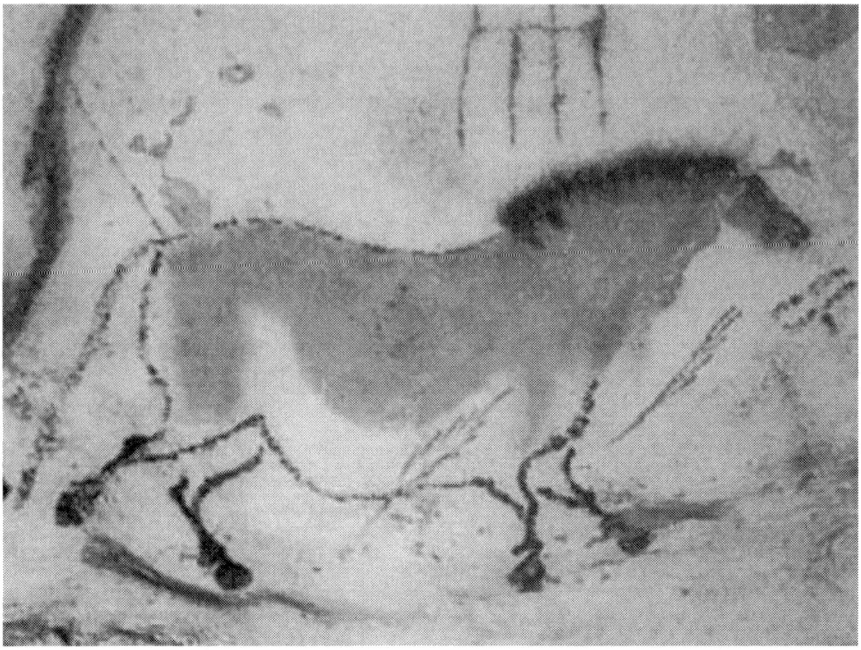

Image of a horse from the Lascaux caves in France.
Credit: Cro-Magnon peoples (Public Domain)
http://commons.wikimedia.org/wiki/File:Lascaux2.jpg

Cave art in Spain. Credit: Rameessos, National Museum and Research Center of Altamira (Public Domain)

ARCHEOLOGY AT UR

Sir Leonard Woolley, in his 1929 landmark book, ***Ur of the Chaldees: A Record of Seven Years of Excavations***, described what he and his team of 400 workers excavated from 48 graves near Ur. In addition to the British Museum, he was also funded by the University of Pennsylvania in Philadelphia. Many of his finds were sent there to be studied and catalogued. One of his major finds sent to Philadelphia was the well preserved skeleton found 41 feet below the surface. The 41 feet of material above the body contained remains of mudbrick walls, pottery, and various Neolithic items. The body had been buried in the top layer of a 12-foot-thick layer of clean water-lain silt. At the bottom of the silt layer were mudbrick walls, flint, and unfired ("green") pottery. In 2012, the Philadelphia Museum dated the skeleton to be 6,500 years old.

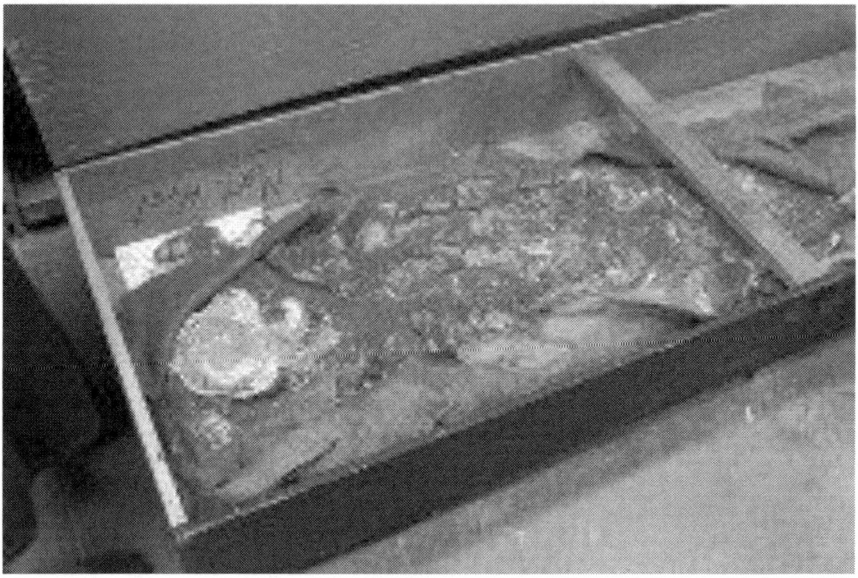

6,500-year-old skeleton (still bound in wax) found in Ur. Credit: University of Pennsylvania Museum of Archaeology and Anthropology

Woolley felt the 12-foot layer of silt was laid during the flood. Below the silt represented life in Ur before the flood. The body was of a person who lived after the flood and was buried into the silt from the flood. The 41 feet of debris was accumulated between the time of the flood and the present. Below is the author's depiction of Woolley's find.

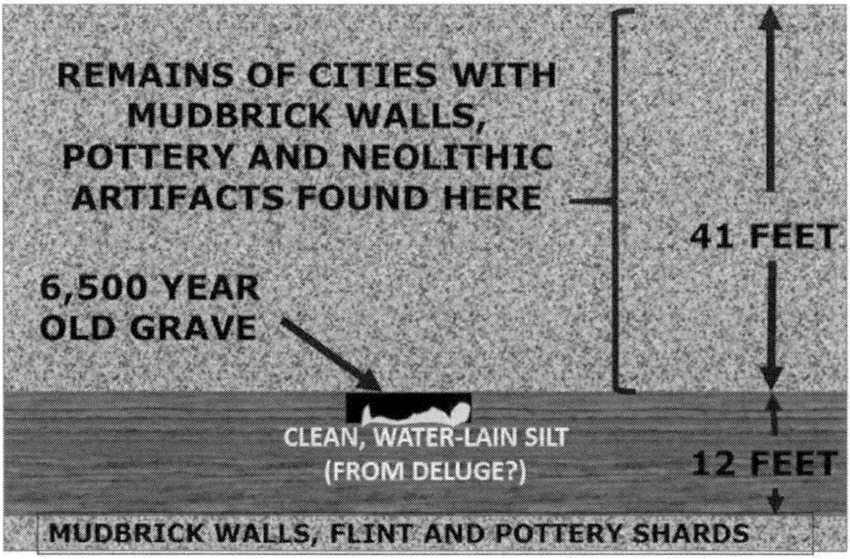

Author's depiction of Woolley's find.

Thus, according to our hypothesis, before the flood the people living at Ur, Urick, Shuruppak, and other Sumerian cities were middle stone-age (Mesolithic or pre-Neolithic) people as evidenced by the flint shards and green pottery. After the flood, humanity eventually returned to the region and had progressed to the bronze age. This formed the beginning of civilization as we know it, resulting in all of the inventions listed at the beginning of this book. A magnificent civilization sprang up capable of building structures like the one shown in the figures below.

Ziggurat (Shrine to moon god) at Ur (temple platform) made of mud bricks and held together with pitch. (Public Domain) Credit: http://commons.wikimedia.org/

Shrine to moon god, Nanna (son of Enlil and Ninlil), from the 1939 reconstruction by Leonard Woolley (Excavations at Ur vol. V, fig. 1.4) Wiki Commons, Public Domain

Chapter 13
After the Deluge

In 1994 the German Archaeological Institute (DAI) began a dig at Göbekli Tepe near Şanlıurfa (Urfa), Turkey, under the direction of the late Professor Doctor Klaus Schmidt. The tell is located at 37° 13.39746′N 38° 55.34633′E. It has a height of 49 feet and a diameter of about 984 feet. Göbekli Tepe is close to the headwaters of the Euphrates River, about six miles from Urfa, and is at the highest point overlooking Harran to the south. Its elevation is 2,500 feet, as compared to Noah's landing site at perhaps only 500 feet.

Material found thus far has been dated to be 12,000 years old. Dr. Schmidt said that after 14 years they had uncovered only about 5 percent. He believed that when they go deeper, they will find the site to be over 14,000 years old. He said villages surrounding this site are of the same time period and were occupied by hunter-gatherers. The villages had large storage facilities for storing wild grains. Perhaps people came here periodically from great distances for religious ceremonies. The location is shown on the map below.

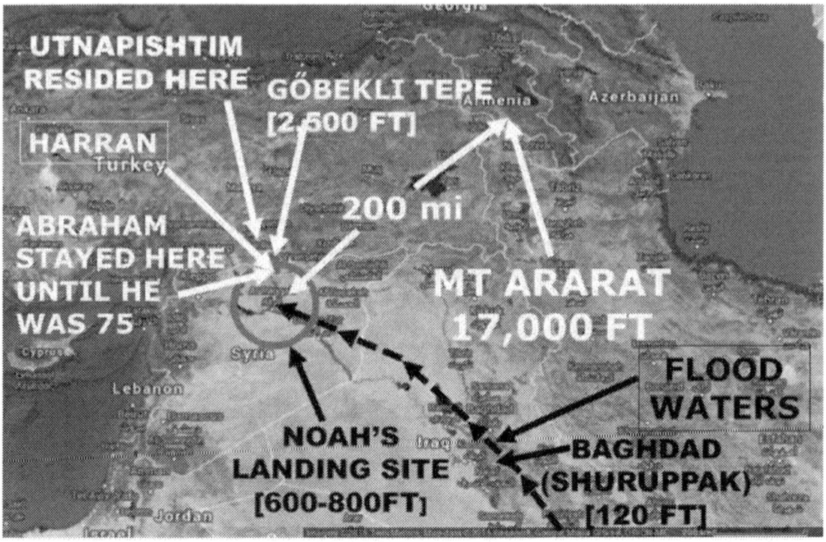

Author's depiction of Göbekli Tepe relative to Noah's landing site. The name Göbekli Tepe means "belly hill" in Turkish. Source map: Google Maps

Below is the dig at level 1 of Göbekli Tepe. There are over 200 pillars, some 20 feet high weighing 20 tons each and inserted into sockets in the bedrock. The columns are arranged in circles of at least ten surrounding two taller columns in the center.

Stone ring structures A – D. Credit: Teomancimit, Göbekli Tepe, Şanlıurfa, Wikimedia Commons

Artist's depiction of three stone rings. Note inner chamber (sanctum?) with at least ten lesser columns surrounding two main (higher) columns. Credit: http://www.ancient-wisdom.co.uk/turkeygobekli.htm

Above is an artist's depiction of one of the temples at Göbekli Tepe. Note that there are two large columns in the center and at least ten lesser ones around the perimeter of the inner ring. According to Sitchin, Sumerian mythology talks about periodic meetings of their pantheon of gods. No meeting could take place without all twelve gods in attendance. The two main gods were Anu and Enlil. Enki and the remaining other nine would be required to attend.

Stanford University anthropologist Ian Holder has made the observation that the pillar carvings are dominated by menacing creatures such as lions, spiders, and snakes. One would expect the carvings to be of edible prey like gazelle, wild boar, and deer which would have been hunted by the hunter-gatherer people who built Göbekli Tepe.

Perhaps the reason for carvings of menacing creatures is that the Sumerians assigned to each of the twelve gods (lords) their own zodiac sign. If Göbekli Tepe was a Sumerian temple to their lords, then scattered around the site and on the columns would be carvings of the zodiac symbols of their lords. See figures below.

Stone carving of wild boar with birds. Credit: http://www.ancient-wisdom.co.uk/turkeygobekli.htm

Stone carving with pelicans and scorpion. Credit: http://www.ancient-wisdom.co.uk/turkeygobekli.htm

"Göbekli Tepe 2" by Teomancimit - Own work. Licensed under CC BY-SA 3.0 via Wikimedia Commons

Stone carving of a lion.
Credit:
http://www.ancientwisdom.co.uk/turkeygobekli.htm

Stone carving of predatory animal on pillar 27 from Enclosure C (Layer III). Credit:Wikimedia Commons

Pillar with sculpture of a fox.
Credit: Zhengan, Wikimedia Commons

Author's depiction of 18 foot pillar. Note anthropomorphic style showing hands and loin cloth. Based on shadows shown in the original photo, the figure is facing south-south-east.

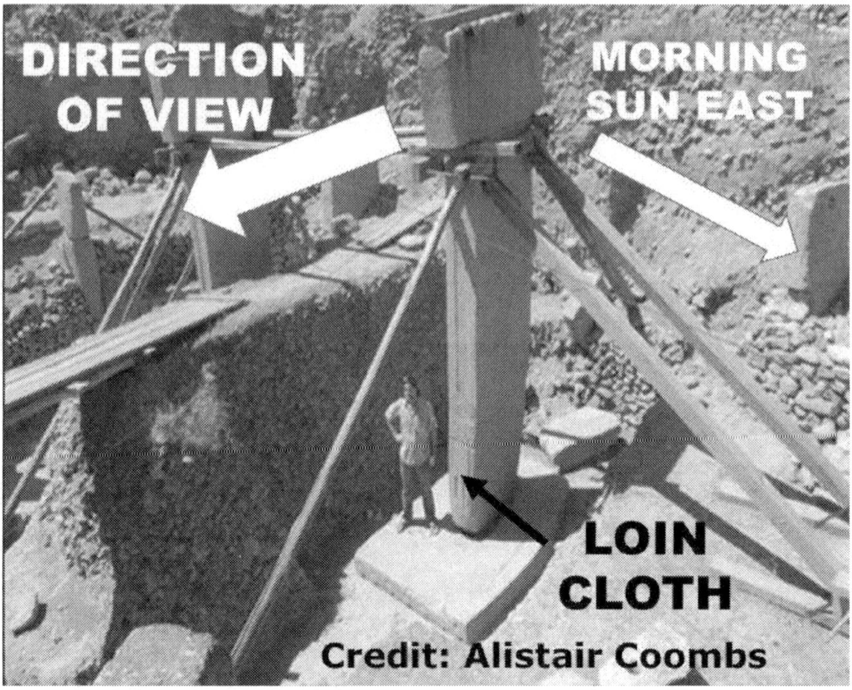

Photo of T-shaped column facing south as determined by shadows. Credit: Alistar Coombs for ancient-origins.net

There are four exposed rings (enclosures) but ground penetrating radar indicates there are sixteen still buried. In all, there are twenty of these enclosures at the Göbekli Tepe site. For some reason, once the stone rings were finished, the ancient builders covered them over with dirt. The hilltop was created over the centuries by layers of buried enclosures, according to Schmidt.

It would be easy to imagine that a ceremonial enclosure was constructed each time the gods (lords) met. 3,600 years was important to them so if they met every 3,600 years, and each level contained four rings then each level would represent 14,400 years. However, archeologists date this level at about 11,500 years old. It is interesting that for some reason the site was deliberately buried about 8,000 years ago. That would be about 3,600 years after the last level was built. No one knows why they were buried. However, this was about the time humanity returned to the Tigris-Euphrates valley and their religious centers would have been moved back to Sumer.

Göbekli Tepe buried 8,000 years ago.
Credit: Wikimedia Commons

Another interesting fact is that for each enclosure, the two central columns in the enclosure rings of inner stones are pointed in a southerly direction. This is evidenced by the shadows in the photo taken by Alistair Coombs of T-shaped pillars at Göbekli Tepe. The pillar is the same 18 foot t-shaped pillar shown above in the author's drawing. According to Sitchin, the Sumerians believed their gods came from the planet Nibiru which would approach from the Earth's southern hemisphere every 3,600 years. The central pillars were arranged to look over the Plain of Harran to greet the arrival of Nibiru as it rose from the southeastern horizon. Below is the author's drawing taken from an aerial view of enclosures A, B, C, and D. The arrows show the direction the two central columns are aligned and facing.

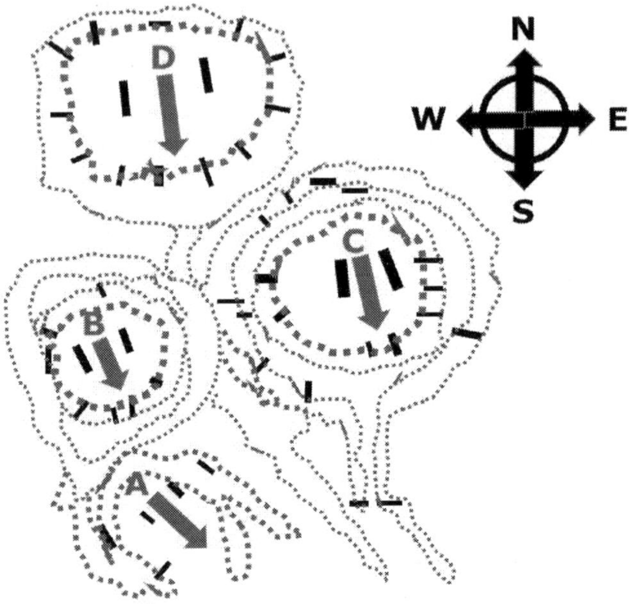

Author's depiction of direction the central columns point in enclosures A, B, C, and D.

It can be seen that the central columns face in the general southerly direction. The fact that they do not all point in exactly the same direction may be because they were not built at the same time and Nibiru would rise from a slightly different direction as the Earth and Nibiru orbited the Sun. Nibiru's period is not exactly an integer of Earth years which creates a slight "phase shift". Assuming a period of 3,600 years and a perigee of 2.7 AU, Nibiru would spend less than four years inside the orbit of Jupiter. It probably would not be visible beyond Jupiter.

It does not end with Göbekli Tepe. A sister site in Karahan Tepe is under development. It covers an area of about 33 acres and is located 22 miles (35.5 km) northeast of the ancient city of Harran, 23 miles (37 km) east-southeast of Göbekli Tepe and 39 miles (63 km) east-southeast of the city of Şanlıurfa (Urfa). Discovered in 1997, it was dated to c. 9500–9000 BC by Turkish archeologist Bahattin Çelik. Thus far his team has conducted two surveys, the first in 2000 and the second in 2011. Excavation work has just begun, but the site contains many T-shaped stone pillars, which are anthropomorphic in nature and bear carvings either in high relief or 3D. Statues of goats, gazelles, and rabbits made out of polished rock have been found in the area.

Another Neolithic site overlooking the Harran plain is Hamzan Tepe which is six miles (ten km) south of Urfa.

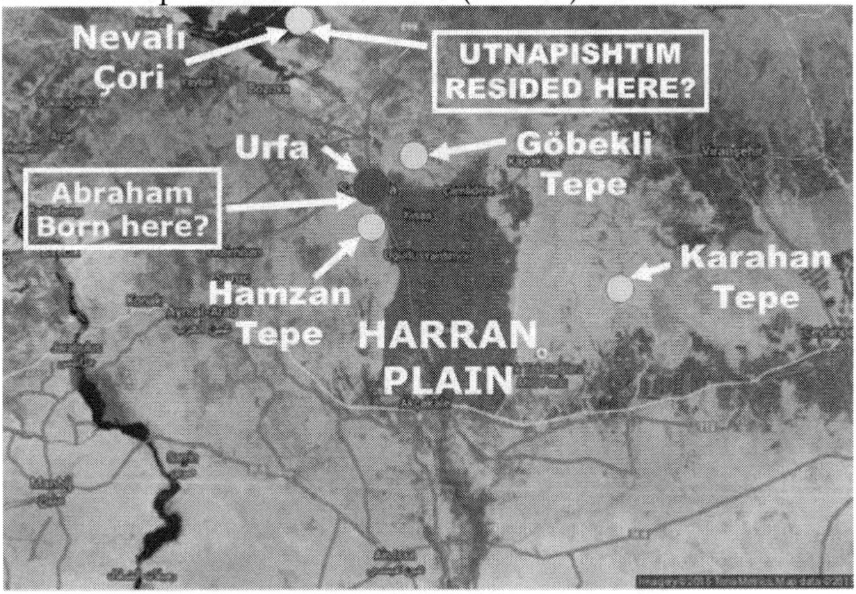

Neolithic sites overlooking the Harran plain.
Source map: Google Maps

Looking at maps and photographs of this region of Turkey around Göbekli Tepe and Harran, one is struck with the arid looking environment with sparse vegetation. It features extremely hot, dry summers and cool, moist winters. It is so hot in the summer that archeologists do not work here. One would wonder what the appeal was for people living here 14,000 years ago. Is this area where humanity re-established itself after the Deluge? Archeologist Schmidt believed that back then this was probably a true "garden of Eden" with plenty of water, vegetation, and wild game.

Recall from the Old Testament that Terah, Abraham's father, took his family there on the way to Canaan but decided to stay in Harran instead. Some think that Abraham was actually born in nearby Urfa (Şanlıurfa) instead of Ur. This would make more sense when looking at the map of Abraham's migrations. Why would Terah divert up to Harran on his way to Canaan? It makes more sense that he would have simply moved down the road from Urfa to Harran on the way to Canaan and stayed there because his son Harran may have died there. It is believed that Terah named the plains after him and lived in Harran for the rest of his life. Harran was Abraham's younger brother by 5 years and was the father of Lot. Before Terah died, Abraham at the age of 75 was called by the Lord to go to Canaan and start a new religion. Abraham then rejected polytheism and took Harran's son, Lot, into Canaan where he founded the Hebrew religion based on only one God.

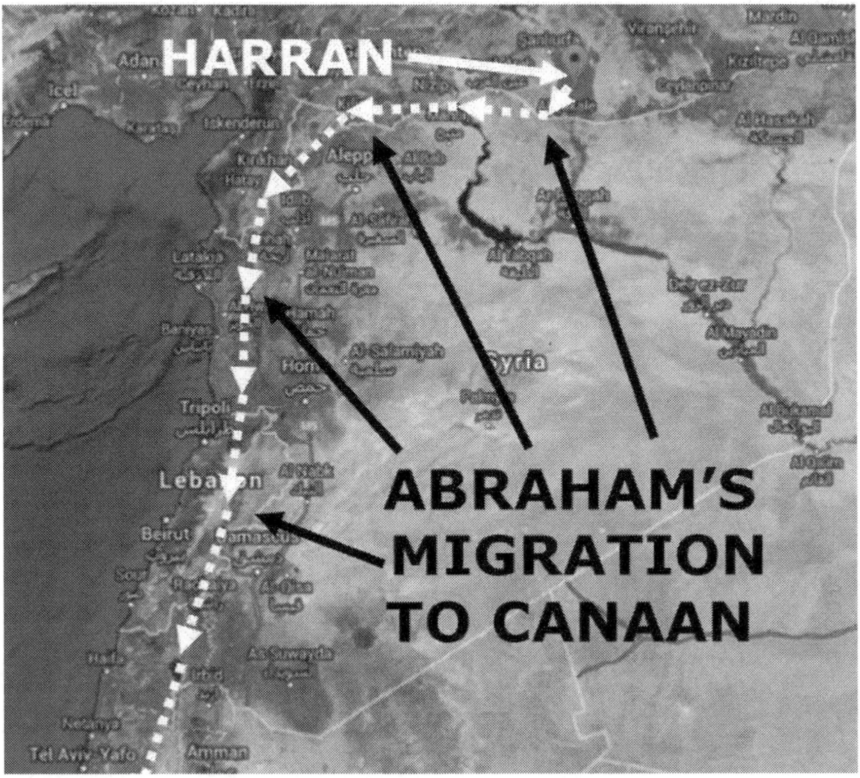

Author's depiction of Abraham's migration from Harran to Canaan. Source map: Google Maps

INSTRUCTION OF ABRAHAM

There is a tradition in Orthodox Judaism that Abraham stayed in the house of Noah and his son Shem where he received instructions and historical knowledge from both. Perhaps it was in the area of Göbekli Tepe, Urfa, and Harran that all of this occurred. One might wonder how that could have occurred with ten generations separating Abraham from Noah. The figure below explains it. Assuming the flood occurred 14,200 years ago, then Abraham lived from 14,010 to 13,835 years ago. Noah and Shem did not die until 13,850 and 13,700 years ago, respectively. Clearly Noah and Shem could have been known by Abraham on a personal basis.

PATRIARCH	BORN (BCE)	DIED	LIVED
NOAH	14,800	13,850	950
SHEM	14,300	13,700	600
ARPACHSHAD	14,299	13,861	438
SHELAH	14,265	13,832	433
EBER	14,235	13,771	464
PELEG	14,201	13,962	239
REU	14,171	13,932	239
SERUG	14,139	13,909	230
NAHOR	14,109	13,961	148
TERAH	14,080	13,875	205
ABRAHAM	14,010	13,835	175

Ten generations between Noah and Abraham. (Genesis)
Note: Noah is alive during Abraham's life

Below is a map showing the migration of Noah's three sons Ham, Shem, and Japheth and their descendants. It assumes Noah's descendants began their migration from the region near Göbekli Tepe.

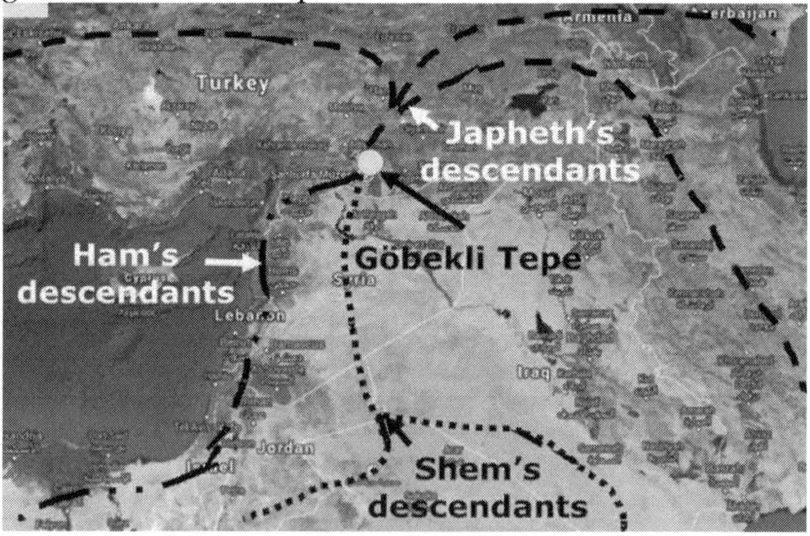

Author's depiction of Noah's descendants' migrations.
Source map: Google Maps

Below is another interesting map which shows the intersection of the three nations of Noah to be at a location close to Göbekli Tepe.

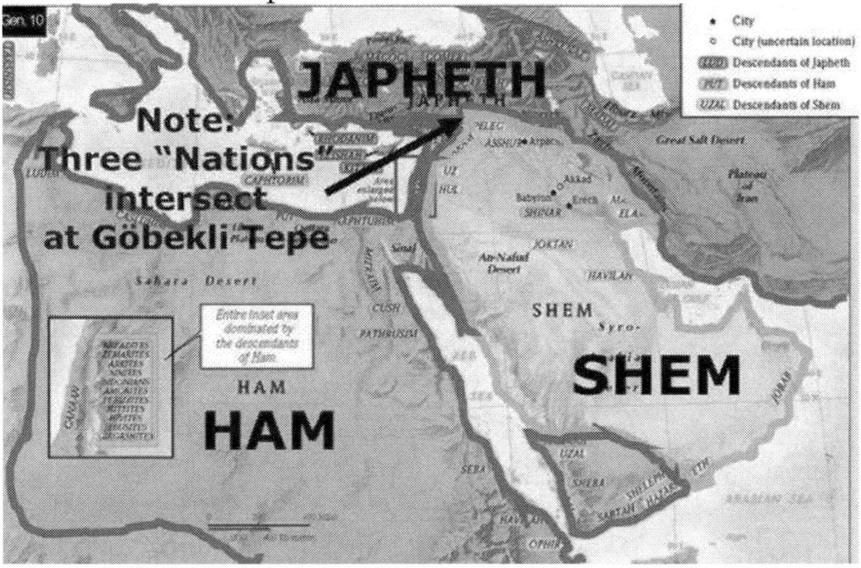

Intersection of three "nations."
Credit: **www.babylonrisingblog.com**

Archeologist Schmidt believed it was the extensive and coordinated effort required to build the monoliths at Göbekli Tepe that laid the foundation for the development of complex societies. He believed the whole innovation process of making the transition from hunting and gathering to turning wild animals into domestic livestock happened at Göbekli Tepe first and then spread south.

SPREAD OF AGRICULTURE

12,000 years ago there were an estimated 5 million people on Earth. Archeological data indicate that at about that time, the domestication of plants and animals evolved at various locations worldwide. Below is a map showing the spread of agriculture in the eastern part of the world starting about 12,000 years ago. Data was extracted from a map at **http://www.maps.com/ref_map.aspx?pid=11437**.

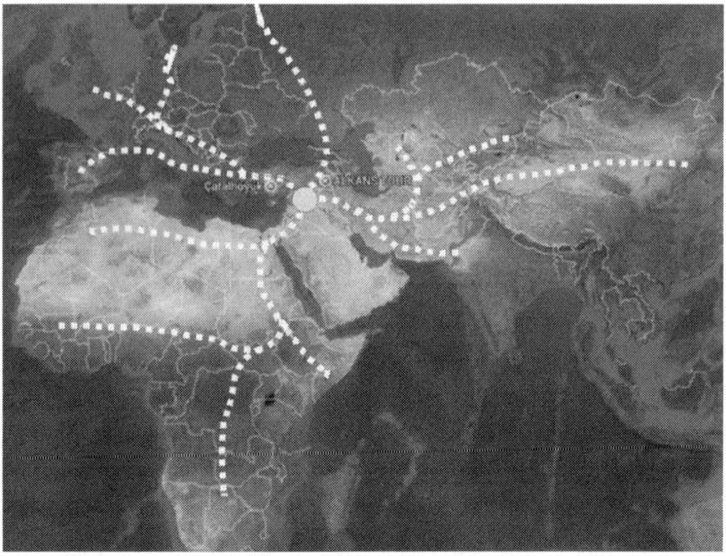

Author's depiction of the spread of agriculture in the eastern part of the world. Source map: Google Maps

Farming appears to have originated from the region near Göbekli Tepe.

Spread of agriculture 12,000 years ago.

Source map: Google Maps

Just 20 miles to the northeast of Göbekli Tepe, at a prehistoric village, geneticists found the world's oldest domesticated strain of wheat dated at 10,500 years ago. Farming spread from there down into Mesopotamia as well as into Europe, Egypt, and the Indus Valley. Crops such as wheat, barley, and lentils were crucial in establishing complex civilizations. At Göbekli Tepe researchers have found large limestone vessels capable of holding up to 160 liters (42 gallons) of liquid. Preliminary chemical tests of the inner lining indicate these vessels may have been used to soak, mash, and ferment grain into beer. Early grain crops may have been far better suited for the production of beer or wine than bread. Converting wheat into bread required developing a suitable wheat.

Along with *EMMER WHEAT, EINKORN* wheat is among the forms of wheat that were first cultivated by humans. See below.

EINKORN WHEAT **EMMER WHEAT**

Credit: Wikimedia Commons

In 1997 Manfred Heun et al., using genetics, located the origin of einkorn wheat to Karaca Dag, Turkey.

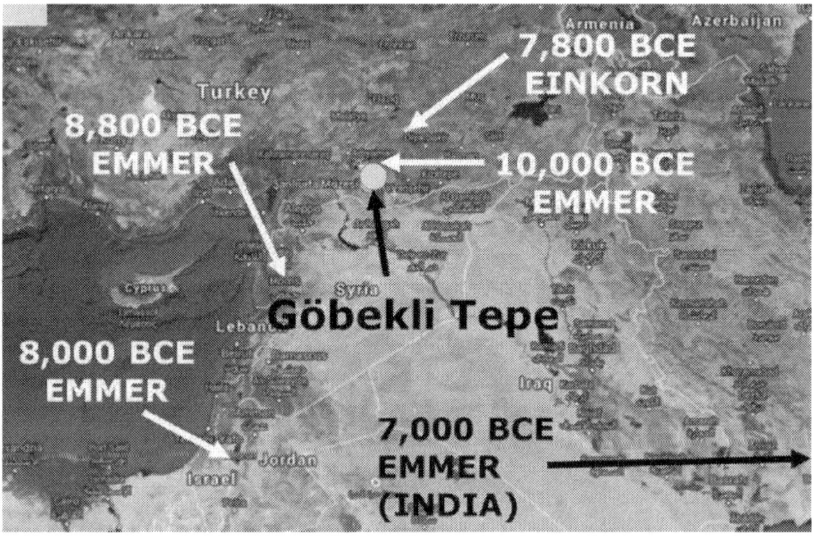

Spread of einkorn and emmer wheat.
Source map: Google Maps

Agriculture spread along the Zagros Mountains east of Sumer into Susa in the proto-Elamite Empire by 7,000 BC. From Susa it spread to Uruk about 4,400 BC.

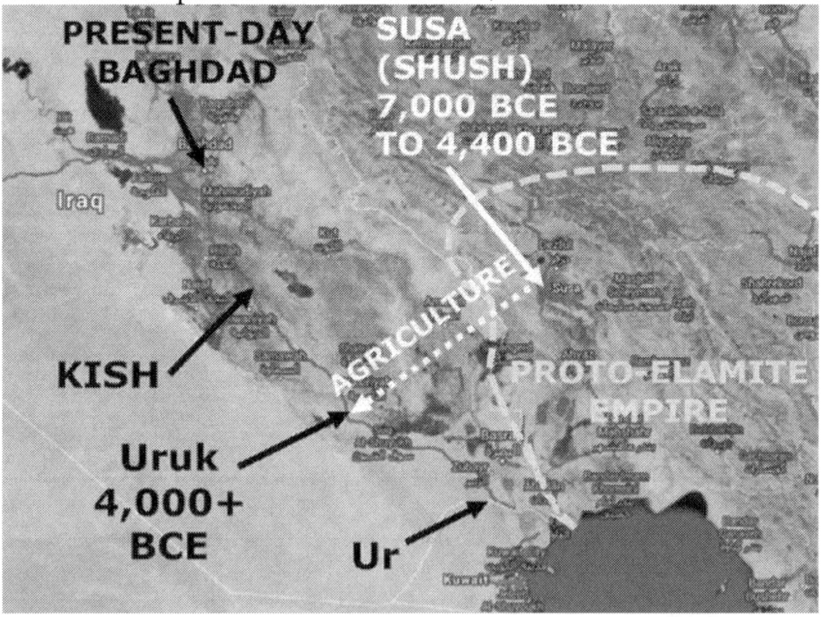

Author's depiction: spread of agriculture from the Elamite territory into Sumer. Source map: Google Maps

According to Sitchin, the mountainous land east of Sumer was called "E. LAM" which means "house where the vegetation germinated." According to Sitchin, legend says Enlil had cereals extended from E. LAM to Sumer. Sitchin translates Sumer to "Sumer, the land that knew not grain, came to know grain."

CIVILIZATION RETURNS TO THE VALLEY

Myth and traditions about the Tigris-Euphrates valley along with periodic surges in sea level may have kept people away.

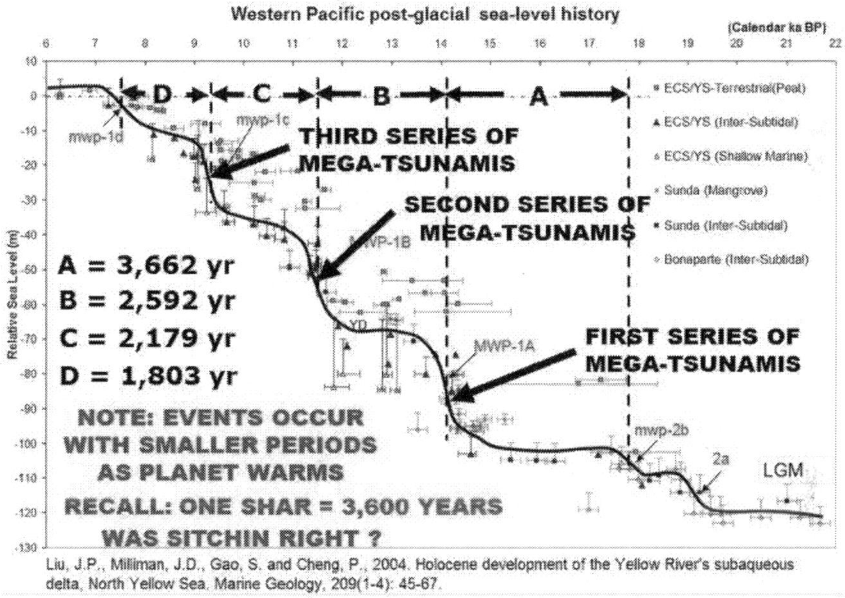

Graph of unstable conditions in western Pacific. Credit: Liu, J.P., Milliman, J.D., Gao, S. and Cheng, P., 2004. Holocene development of the Yellow River's subaqueous delta, North Yellow Sea. *Marine Geology*, 209(1–4): 45–67

The graph above shows what may have been a series of tsunamis. Significant rises in sea level may provide evidence that tsunamis occurred about 14,200 years ago as well as 11,500 and 9,300 years ago. Note that the period between tsunamis is not a constant 3,600 years as Sitchin would have expected if the cause was the periodic approach of Nibiru. What the decreasing period seems to indicate is that as the planet warmed, the melting process was faster and the amount of ice falling into the ocean was decreasing. Below is a map showing how re-occurring Antarctic mega-tsunamis would have aligned with chevrons found around Indian Ocean.

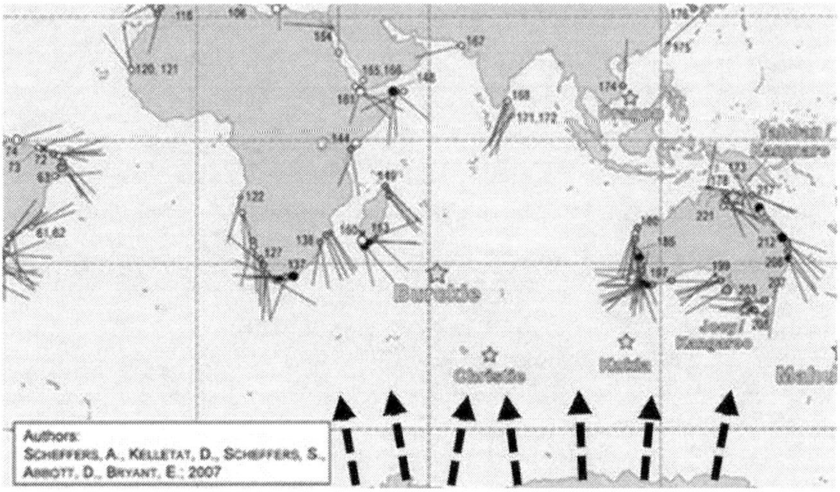

Author's depiction of tsunamis from Antarctica relative to alignment of chevrons.
Source map: Holocene Impact Working Group
http://tsun.sscc.ru/hiwg/chevrons.html

Perhaps the twelve feet of clean water-lain silt Woolley found forty-one feet down at Ur was from the tsunami which occurred 9,500 years ago and not from the main event (Noah's Flood) 14,200 years ago. In a perfect world, an archeologist reading this book in the future may have the opportunity to return to Ur and dig deeper.

During this time when the Antarctic ice sheets were still unstable, these recurring tsunamis may have plagued the valley and kept humanity away. Eventually, things settled down, and humanity did return to the valley. This is evidenced by the spread of agriculture. Agriculture in Sumer was the key to the rise of the world's first civilization, the invention of writing, and the beginning of recorded history.

Conclusions

What surfaced from all of the turmoil that has been described here was nothing more than the beginning of civilization. All of the civilizations in recorded history can trace their roots back to Sumer. Their history, both written and transmitted through ancient myths, tells us that a flood event occurred that was so dramatic that it separated prehistory from recorded history and the dawn of civilization. There is little doubt as to where the flood occurred, but much work lies ahead to determine when and how it occurred. For those who think they know the answer, I offer the following quote: **"The ultimate barrier to truth is the conviction we already have it!"** (author: Chuck Missler).

Based on the information presented in this book, the following scenario of events is offered for consideration. Consider that 14,200 years ago the population of Homo sapiens was growing in the southern region of Mesopotamia. These were Paleolithic hunter-gatherer people with no ability to record historical events other than through oral traditions and folktales.

During the exit from the last global ice age maximum, beginning 20,000 years ago, a dramatic warming period occurred which resulted in the collapse of huge mile-high ice sheets. As they plummeted into the Indian Ocean, unprecedented mega-tsunamis invaded the lake we now call the Persian Gulf and traveled up the Tigris-Euphrates valley. All life was drowned except for a few humans (Noah's family) who survived and were carried northwest up the valley and deposited near the present plain of Harran. As the waters began to recede, the survivors (Noah's family) settled into the northwest end of the plain of Harran and founded Urfa. At that time, the climate was well suited for their survival with plenty of vegetation and game to keep them fed.

They were religious (polytheistic) people, and as their population grew, they had the manpower and skills to build temples to their pantheon of twelve gods. These temples were at Göbekli Tepe, Hamzan, and Karahan and overlooked the plain of Harran. Eventually, agriculture was developed and the population began to spread east and west of the plain of Harran, staying mostly in the highlands. The descendants of Ham migrated into Canaan and down into Africa. The descendants of Japheth spread west into the Aegean Sea area and southern Europe as well as east along the Zagros Mountains into India. The descendants of Shem migrated into the Arabian Peninsula and eventually into the Tigris-Euphrates valley.

Stories of the flood catastrophe and the continued threat of repeated tsunamis persisted for generations and kept them out of the valley.

By seven thousand years after the flood, the descendants of the original survivors (Noah's family) extended as far west as the Mediterranean Sea and as far east as the town of Susa in the Zagros Mountains which were located to the east of the Tigris-Euphrates valley. The climate became more arid and the threat of tsunamis was gone so people returned to the sites of their ancient towns in Sumer such as Ur, Uruk, and Kish. With them came agriculture, and with abundant water and irrigation, civilization flourished. Their religious temples overlooking the plain of Harran were shut down, buried, and transferred to new sites in Sumer. The invention of writing finally allowed them to record the ancient myths passed down over the millennia, giving us a glimpse of their prehistory. The rest, as they say, is history.

For readers who want to witness Noah's Ark firsthand, they can find it in Hong Kong. The park opened in May of 2009 after seventeen years of planning. The project was funded by the Kwok brothers and five Christian organizations. The ark is about 300 cubits (450 feet) long and 50 cubits (75 feet) wide, just as the Lord prescribed.

Hong Kong theme park in 2009.
Credit: HK Arun, Wikimedia Commons

Appendix A
Solar insolation

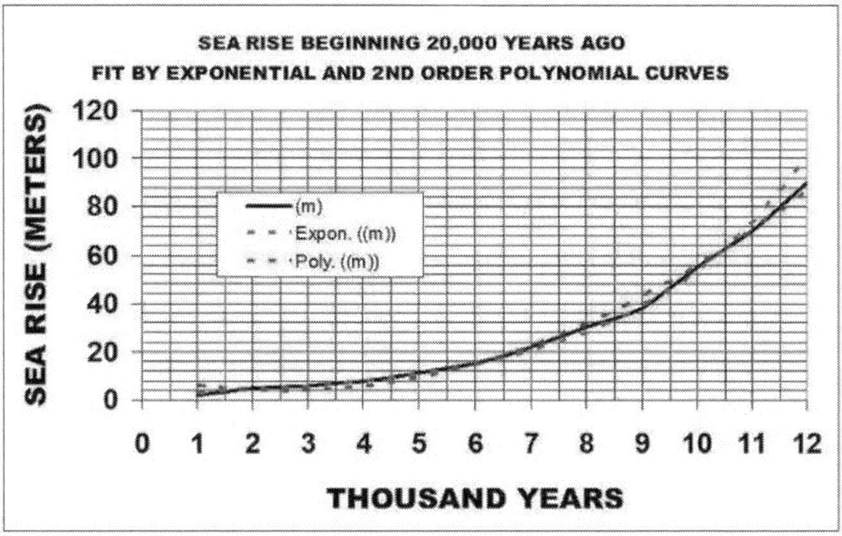

Author's fit of sea rise data.

The change in slope of the curve as sea level rises can be explained by three factors: increase in solar insolation, reduced albedo as melting ice exposes more land, and increased surface area of the world's oceans. The figure below demonstrates the latter point.

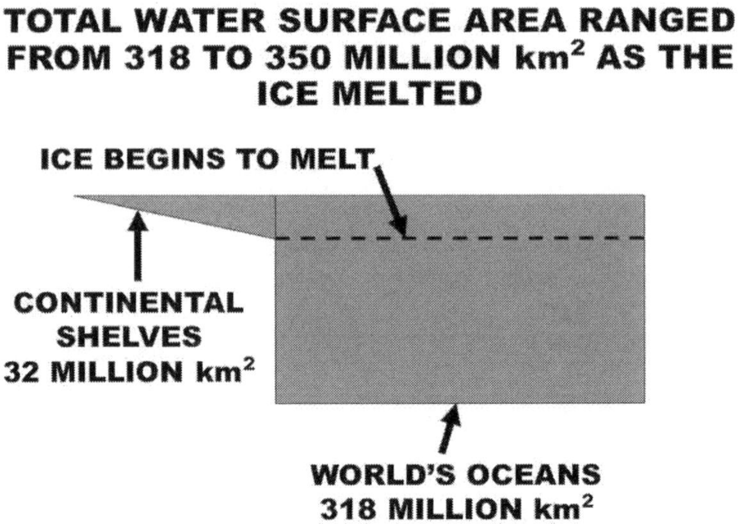

Author's illustration of increase in ocean surface area as continental shelves are flooded from sea rise.

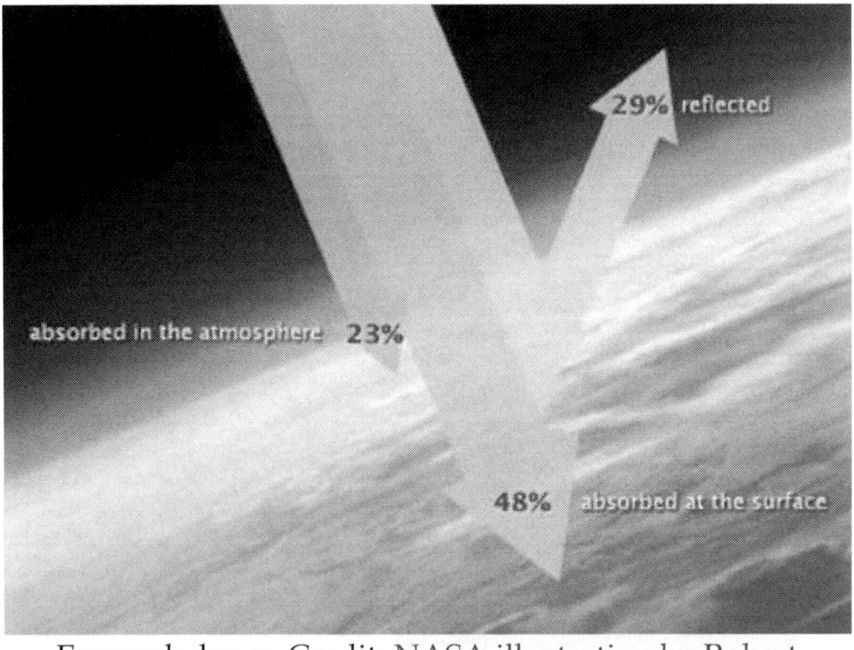

Energy balance. Credit: NASA illustration by Robert Simmon. Astronaut photograph ISS013-E-8948

There are approximately 340 watts per square meter of solar energy striking the Earth. The figure above shows that 29 percent is reflected back into space primarily by clouds and bright objects such as snow and ice. The atmospheric gases and particles in the atmosphere absorb about 23 percent. The remaining 48 percent is absorbed by the Earth's surface.

In the graph below, the author used the rising sea level and heat of fusion to estimate the net solar energy absorbed.

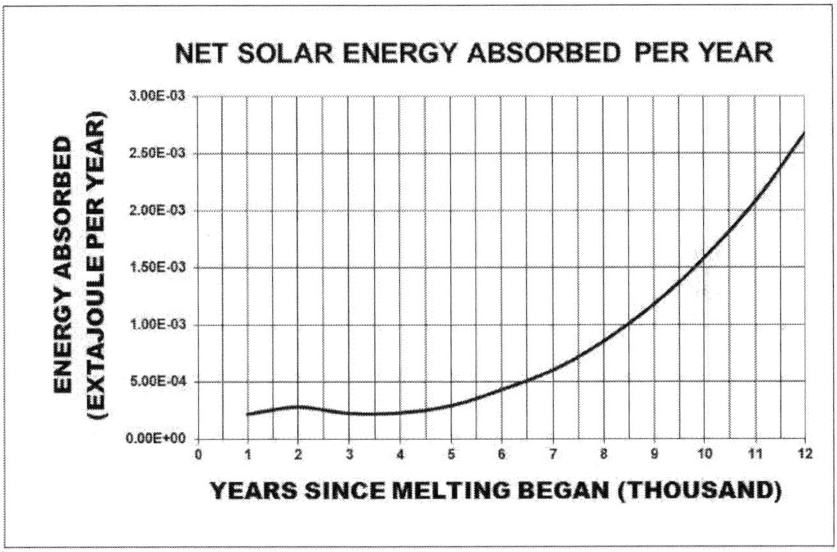

Net solar energy absorbed by melting ice/rising sea level.
Credit: author

- Note: Approximately 3.2 extajoules (10^{18}) is required to melt enough ice to raise the oceans about 25 meters (82 ft).
- It took nearly 5,000 years to melt that much ice as we came out of the ice age.
- The Earth receives about 0.174 EJ with about 30 percent (0.054 EJ) reflected back to space.
- With so much ice covering the planet, perhaps only 0.10 EJ reaches the surface per year.

- 15,000 years ago about .0003 EJ per year was melting ice.
- To raise ocean levels 25 meters in one year would require over 3 EJ or 30 times the total annual solar energy reaching the Earth's surface.
- Total area of the oceans today is 350 million km^2.
- Total area of the continental shelves is 32 million km^2.

Appendix B
Nibiru's gravitational influence

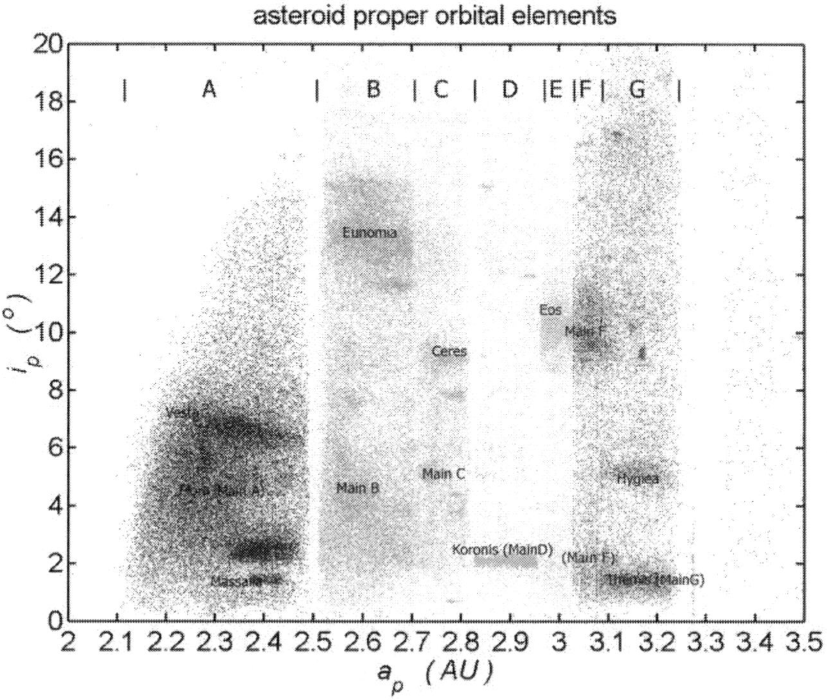

Plot of proper inclination vs. semi-major axis for numbered asteroids. (Wikimedia Commons)

Other data, not shown on the graph above, indicates a significant number of asteroids have orbital planes tipped as much as 30 degrees. It is assumed that Nibiru (Planet X) has its closest approach at the asteroid belt. Using the plot above, a value of 2.7 AU is assumed. Using the governing equation given below, the gravitational influence can be compared with that of the Moon.

$$F = G\,(m_1 \times m_2)/\,R^2$$

Where: G = the gravitational constant

m_1 and m_2 are masses of two bodies

R = radius (AU) between centers

ASSUMPTIONS:

1. Nibiru's mass is equal to Jupiter's and equals 318 Earths.
2. Distance between Nibiru and Earth is 2.7 - 1.0 or 1.7 AU.
3. Moon's mass equals 0.0123 Earths.
4. Mean distance between the Moon and Earth ranges from 0.00242 to 0.00272 AU with an average being equal to .00257 AU.

If we look at $\mathbf{F_X / F_{Moon}}$, G and the mass of Earth drops out and we get $\mathbf{F_X = F_{Moon}\ (0.00257/1.7)^2 \times (m_x / m_{Moon})}$. This becomes $\mathbf{F_X = F_{Moon}\ (0.00257/1.7)^2 \times (318 / 0.0123)}$. Thus $\mathbf{F_X = 0.0591 \times F_{Moon}}$ which represents less than 6 percent of that from the Moon.

If we compare the force from the Moon at closest approach to the farthest distance, we get $\mathbf{F_{CLOSE} / F_{FAR} = (272/242)^2 = 1.263}$ which represents a total range of about 26 percent.

Thus, the 6 percent attraction from Nibiru falls within the normal range of 26 percent which we normally see and probably would not cause any unusual effects on the Earth.

Bibliography

For the reader whose curiosity is piqued by this book, I have included a list of references that will be helpful for doing more research. They are organized by chapter. Extensive research was done through Google, Bing, and Wikipedia. Those links are given by chapter in the second half of the bibliography.

PRINT REFERENCES

Introduction

Woolley, Leonard C., *Ur of the Chaldees: A Record of Seven Years of Excavations*, 2nd Ed., Ernest Benn Limited, London, 1950

Book of Genesis Chapter 6, King James Version

Chapter 1

Kramer, Samuel Noah, *The Sumerians: Their History, Culture, and Character*, The University of Chicago Press, Chicago & London, 1971

Sitchin, Zecharia, *The 12th Planet: Book I of the Earth Cronicles*, Avon Books, New York, 1976

Chapter 2

Woolley, C. L., *The Sumerians,* W. W. Norton & Company, New York, 1965

Sitchin, Z., *Genesis Revisited,* Avon Books, New York, 1990

Rosen, Megan, "Digging into Europa: The ice of a distant moon. Piercing Europa's frigid shell to search for life below," *Science News*, May 2, 2014

Harrington, Robert S., "The Location of Planet X," Session 1.2, 19th Meeting of the American Astronomical Society, Gaithersburg, MD, 25–26 July 1988

Harrington, Robert S., "The Location of Planet X," *The Astronomical Journal*, Vol. 96, No. 4, October 1988

Crocket, Christopher, "A distant planet may lurk far beyond Neptune," *Science News*, November 14, 2014

Wilford, John N., "Looking for Planet X: Old Clues, New Theory," *New York Times*, July 1, 1987

Gladman, Brett J., et al., "On the asteroid belt's orbital and size distribution," *Icarus* 202 (2009) 104–118

Yeager, Ashley, "Rosetta casts doubt on comets as Earth's water providers: Comet 67P's atmosphere contains a surprisingly high fraction of deuterium," *Science News*, December 10, 2014

Alexeyev, Mikhail F., LeDoux, Susan P., and Wilson, Glenn L., "Mitochondrial DNA and aging," The Biochemical Society, *Clinical Science* 107 (2004) 355–364

Cortopassi, G. A., Shibata, D., Soong, N. W., and Arnheim, N., "A pattern of accumulation of a somatic delition of mitochondrial DNA in aging human tissue," Proceedings of the National Academy of Science, *Genetics*, Vol. 89, August 1992, 7370–7374

Mutwa, Vusamazulu Credo, *Indaba, My Children*, Grove Press, New York, 1964

Chapter 3

Lambert, W. G., and Millard, A. R., *Atra-Hasis: The Babylonian Story of the Flood*, Clarendon Press, Oxford, England, 1969

Chapter 4

Sanders, N. K., *The Epic of Gilgamesh*, Penguin Books, New York, 1972

Dalley, Stephanie, *Myths from Mesopotamia: Creation, the Flood, Gilgamesh, and Others*, Oxford University Press Inc., New York, 2000

Mitchell, Stephen, *Gilgamesh: A New English Version*, Free Press Div., Simon & Schuster, New York, 2004

Chapter 5

Abram, Simon (editor), *The Holy Bible, Genesis Chapter 6,* King James Version, Kindle Edition, 2013

Chapter 6

White, John, *Pole Shift*, A.R.E. Press, Virginia Beach, Virginia, 13th Printing, 1996

Hapgood, Charles H., *Earth's Shifting Crust* (Foreword by Albert Einstein), Pantheon Books, Inc., New York, 1958

Hapgood, Charles H., *The Path of the Pole: Cataclysmic Poleshift Geology*, Adventures Unlimited Press, Kempton, Illinois, 1999

Brown, Hugh Auchincloss, *Cataclysms of the Earth*, Twayne Publishers, Inc., New York, 1967

Chapter 7

Ryan, William, Ph.D., and Pitman, Walter, Ph.D., *Noah's Flood: The New Scientific Discoveries About the Event that Changed History*, Touchstone, New York, 2000

Scarborough, Robert, "Climate Change and Sea Level Rise," Delaware's Sea Level Rise Initiative, *Delaware Coastal Programs*, Dover, 2009

Chapter 8

Kelletat, Dieter, and Scheffers, Anja, "Chevron-Shaped Accumulations Along the Coastlines of Australia as Potential Tsunami Evidences?", *Science of Tsunami Hazards*, Volume 21, Number 3 (2003) 174

Abbott, Dallas, Bryant, Edward A., Gusiakov, Vacheslev, and Masse, W. Bruce, "Report of International Tsunami Expedition to Madagascar August 28–September 12, 2006," Lamont-Doherty Earth Observatory of Columbia University, New York, posted *International Tsunami Bulletin Board*, 2006

Masse, W. Bruce, "The Archeology and Anthropology of Quarternary Period Cosmic Impacts," Chapter 2 in *Comet/Asteroid Impacts and Human Society: An Interdisciplinary Approach*, edited by Peter T. Bobrowsky and Hans Richman, Springer, Berlin, Heidelberg, New York, 2007

Abbott, Dallas, "Chevron Dunes in Madagascar: The Most Spectacular Tsunami Deposits on Earth," IEEE Aerospace Conference Proceedings: 1 (2007) DOI:10.1109/AERO.2007.352678

Pinter, Nicholas, and Ishman, Scott E., "Impacts, mega-tsunami, and other extraordinary claims," *GSA Today*, January, 2008

Abbott, D. H., et al., "Odd $CaCO_3$ from the Southwest Indian Ocean near Burckle Crater Candidate: Impact Ejecta or Hydrothermal Precipitate?", *40 Lunar and Planetary Science Conference* (2009)

Jackson, Kelly L., et al., *Holocene Indian Ocean tsunami history in Sri Lanka,* The Geological Society of America, GAS Data Repository 2014305

Chapter 9

Gusiakov, Viacheslav, Abbott, Dallas H., Bryant, Edward A., Masse, W. Bruce, Breger, Dee, "Mega Tsunami of the World Oceans: Chevron Dune Formation, Micro-Ejecta, and Rapid Climate Change as the Evidence of Recent Oceanic Bolide Impacts," *Geophysical Hazards International Year of Planet Earth*, Springer, Berlin, Heidelberg, New York, 2010, 197–227

Chapter 10

Imbrie, John, and Imbrie, Katherine, *Ice Ages: Solving the Mystery,* Harvard University Press, Cambridge, Massachusetts, 1979

Chapter 11

Leakey, R., *The Origin of Humankind,* Basic Books, New York, 1994

Berger, L., with Hilton-Barber, B., *In the Footsteps of Eve: The Mystery of Human Origins,* National Geographic Adventure Press, Washington, DC, 2000

Soares, Pedro, et al. "Correcting for Purifying Selection: An Improved Human Mitochondrial Molecular Clock," *The American Journal of Human Genetics,* Volume 84, Issue 6 (4 June 2009) 740–759

Chapter 12

McKie, Robin (Science Editor), "Stone Age sailors 'beat Columbus to America'," *The Observer*, November 28, 1999

Chapter 13

Heun, Manfred, Basilio, Borghi, and Salamini, Francesco, "Einkorn wheat domestication site mapped by DNA fingerprinting," www.dlib.si/stream/URN:NBN:SI:DOC-786RAUGS/644df458.../PDF , 1998

Bower, Bruce, "Agriculture's roots spread east to Iran: Finds at ancient village extend early crop cultivation across the Fertile Crescent," *Science News*, August 24, 2013

Brennan, H., *The Secret History of Ancient Egypt*, Berkley Publishing Group, New York, 2000

Çelik, Bahattin, "Hamzan Tepe in the light of new finds," *Documenta Praehistorica* XXXVII, 2010

WEBSITE REFERENCES

The following are websites referenced by chapter.

Introduction
http://home.earthlink.net/~misaak/floods.htm

www.talkorigins.org/faqs/flood-myths.html
(*Global Flood Stories*, Mark Isaak)

http://en.wikipedia.org/wiki/Leonard_Woolley

http://en.wikipedia.org/wiki/Pyramid_of_Djoser

http://www.kingjamesbibleonline.org/Genesis-Chapter-6/ King James Bible

http://en.wikipedia.org/wiki/Noah_in_Islam

Chapter 1
http://en.wikipedia.org/wiki/Mesopotamia

http://ncse.com/cej/8/2/flood-mesopotamian-archaeological-evidence

http://commons.wikimedia.org/wiki/File:N-Mesopotamia_and_Syria_english.svg

www.bible-history.com

http://en.wikipedia.org/wiki/Royal_Cemetery_at_Ur

http://history.ninjaflower.com/mesopotamia.html

Chapter 2
http://www.sitchin.com/

http://en.wikipedia.org/wiki/Zecharia_Sitchin

http://en.wikipedia.org/wiki/Robert_Sutton_Harrington

http://ad.usno.navy.mil/wds/history/harrington.html

http://en.wikipedia.org/wiki/CFBDSIR_2149-0403 (rogue planet)

http://en.wikipedia.org/wiki/Abraham

http://en.wikipedia.org/wiki/Pluto

http://en.wikipedia.org/wiki/Asteroid_belt

http://en.wikipedia.org/wiki/Main_asteroid_belt

http://burro.astr.cwru.edu/stu/advanced/asteroid.html

https://www.sciencenews.org/article/water-arrived-earth-earlier-thought

https//www.viewzone.com

http://smithsonianscience.org/2012/03/ice-age-mariners-from-europe-were-the-first-people-to-reach-north-america/

http://en.wikipedia.org/wiki/Giant_impact_hypothesis

http://phoenicia.org/zimbabwe.html (160,000 year old bones)

http://www.medicaldaily.com/mitochondrial-eve-common-ancestor-all-humans-actually-lived-160000-years-ago-244742

Chapter 3

http://en.wikipedia.org/wiki/Atra-Hasis

Chapter 4

http://en.wikipedia.org/wiki/Sumerian_creation_myth

http://www.sparknotes.com/lit/gilgamesh/section9.rhtml

http://www.geocreationism.com/history/dating-the-flood-the-epic-of-gilgamesh.html

Chapter 5

http://en.wikipedia.org/wiki/Genesis_flood_narrative

Chapter 6

http://www.physics.org/facts/frog-magnetic-field.asp

http://en.wikipedia.org/wiki/Pole_shift_hypothesis

Chapter 7

http://www.eurekalert.org/pub_releases/2007-11/uoe-fk111507.php

http://www.ldeo.columbia.edu/

http://en.wikipedia.org/wiki/Laurentide_ice_sheet

Chapter 8

http://en.wikipedia.org/wiki/Burckle_Crater

http://en.wikipedia.org/wiki/Holocene_Impact_Working_Group

http://tsun.sscc.ru/hiwg/chevrons.htm

http://www.geocreationism.com/science/dating-the-flood-burckle-crater.html

http://discovermagazine.com/2007/nov/did-a-comet-cause-the-great-flood

http://www.earthmagazine.org/article/giant-dunes-not-mega-tsunami-deposits

Chapter 9

http://en.wikipedia.org/wiki/Megatsunami

http://walrus.wr.usgs.gov/research/projects/tsunamihaz.html

Chapter 10

http://people.rses.anu.edu.au/lambeck_k/pdf/200.pdf (Over 265 papers by Dr. Kurt Lambeck)

http://www.climatedata.info/Forcing/Forcing/milankovitchcycles.html

http://post-glacial sea level.png

http://geology.com/sea-level-rise/florida.shtml. (Interactive sea level program)

http://en.wikipedia.org/wiki/Glacier

http://www.antarcticglaciers.org/antarctica/west-antarctic-ice-sheet/

http://theinconvenientskeptic.com/2013/01/sea-level-and-insolation/

---

http://news.bbc.co.uk/2/hi/south_asia/1768109.stm (Sunken city off of west coast of India)

Chapter 11

http://www.sciencedirect.com/science/article/pii/S0005272806000582

http://www.jci.org/articles/view/64125
(mitochondrial DNA and aging)

http://instaar.colorado.edu/people/john-t-hollin/

http://creation.com/the-date-of-noahs-flood

http://today.uconn.edu/blog/2012/09/symposium-to-highlight-research-partnership-with-jackson-labs/
(Genome Research Research at UCON)

Chapter 12

http://en.wikipedia.org/wiki/Pre-Pottery_Neolithic_A

http://en.wikipedia.org/wiki/Pre-Pottery_Neolithic_B

https://www.google.com/?gws_rd=ssl#q=Ubaid%20pottery

http://ironlight.wordpress.com/2010/03/19/stone-age-sailors-reached-america-18000-years-ago/

http://smithsonianscience.org/2012/03/ice-age-mariners-from-europe-were-the-first-people-to-reach-north-america/

Chapter 13

http://www.smithsonianmag.com/history-archaeology/30706129.html

http://nypost.com/2014/08/21/neanderthals-modern-humans-interacted-for-thousands-of-years/

http://www.smithsonianmag.com/history/gobekli-tepe-the-worlds-first-temple-83613665/#QMSfoLG8YtEv8k5m.99

http://www.bibleplaces.com/haran.htm

http://www.andrewcollins.com/page/articles/Karahan.htm

http://en.wikipedia.org/wiki/Mehrgarh

http://en.wikipedia.org/wiki/%C5%9Eanl%C4%B1urfa

http://www.youtube/JGaGH2WY5Wc (10,500 year-old Karahan Tepe: Göbekli Tepe's Sister Site in Turkey)

http://www.youtube/CXoCRP3isAM (The Unfinished Monolith at Karahan Tepe, Turkey 8,500 BC)

http://en.wikipedia.org/wiki/%C3%87ay%C3%B6n%C3%BC

http://en.wikipedia.org/wiki/Neval%C4%B1_%C3%87ori

http://en.wikipedia.org/wiki/Sin (mythology)

http://en.wikipedia.org/wiki/Neolithic_Revolution

http://leilan.yale.edu/works/akkadian_empire/index.html

http://www.youtube.com/watch?v=rou-_pXVBcA
(The Cave Where Prophet Abraham Was Born, Urfa - Southeastern Turkey)

http://www.cryaloud.com/abraham_abram_susanna_austrian_chronicle.htm

https://www.biblegateway.com/passage/?search=Acts+7%3A2-3&version=ESV

https://www.biblegateway.com/passage/?search=Acts+7%3A2-3&version=KJV

http://www.andrewcollins.com/page/articles/Gobekli.htm
(GOBEKLI TEPE ITS COSMIC BLUEPRINT REVEALED)

http://www.waa.ox.ac.uk/XDB/tours/mesopotamia4.asp

About the author

Dr. Robert E. Farrell has been involved in technology most of his adult life. He holds a B.Sc. in Mechanical Engineering, a Master's Degree in Business Administration, and a Doctor of Engineering. For twenty years, he worked in industry, designing plastics processing machinery. For fifteen years, he was a Professor of Engineering, preparing students for the plastics industry. Dr. Farrell retired with Emeritus status.

During his entire professional career, Dr. Farrell has been fascinated by the UFO phenomenon and his research has led him into the early history of humanity and in particular, the story of Noah's Flood.

This book is the result of an effort to enlighten the reader. He would greatly appreciate any feedback you are willing to give as he continues to research and write more books in both ***The Science Behind...*** series and the ***Alien Log*** series of novels.

Please visit my website: www.alienlog.com

Other books by Robert E. Farrell

Alien Log	(Paperback)	ISBN: 978-0-9759116-0-0
	(E-book)	ISBN: 978-0-9759116-4-8
Alien Log II: The New World Order		
	(Paperback)	ISBN: 978-0-9759116-3-1
	(E-book)	ISBN: 978-0-9759116-5-5
Alien Log III: The Dulce Affair		
	(Paperback)	ISBN: 978-0-9759116-7-9
	(E-book)	ISBN: 978-0-9759116-8-6
Alien Log IV: The Antarctica Affair		
	(Paperback)	ISBN: 978-0-9963587-3-6
	(E-book)	ISBN: 978-0-9963587-2-9
The Science Behind Noah's Flood, 2nd. Ed.		
	(Paperback)	ISBN: 978-0-9759116-9-3
	(E-book)	ISBN: 978-0-9963587-0-5
The Science Behind Alien Encounters, 2nd.Ed.		
	(Paperback)	ISBN: 978-0-9759116-6-2
	(E-book)	ISBN: 978-0-9759116-1-7
	(Hard Cover)	ISBN: 978-0-9963587-7-4

Coming soon

The Science Behind Creation the Universe WITHOUT a Big Bang
(paper and E-book)

The Science Behind Gravitational Field Propulsion: Key to Interstellar Travel
(paper and E-book)

Made in the USA
Monee, IL
10 November 2026

9d0c1e6a-13fe-4d52-8586-288868f72087R01